Sustainability Engineering for Enhanced Process Design and Manufacturing Profitability

Jeffery P. Perl

Sustainability Engineering for Enhanced Process Design and Manufacturing Profitability

Balancing the Environment through Renewable Resources

Second Edition

Springer

Jeffery P. Perl
Private Practice
Lincolnwood, IL, USA

ISBN 978-3-031-52362-5   ISBN 978-3-031-52363-2 (eBook)
https://doi.org/10.1007/978-3-031-52363-2

This Springer imprint is published by the registered company Springer Nature Switzerland AG
The registered company address is: Gewerbestrasse 11, 6330 Cham, Switzerland

Preface

Here is the punch line for this book: A highly developed and largely successful CPI design methodology already exists today for ready application. Its present form is the result of a step/jump improvement brought about by the ESOH regulations of the 1970s. However, we are at a plateau now and ready to ascend to higher levels of continuous improvement through Sustainability Engineering (SE). SE is not just superimposed on top of classic process design but becomes a fully integrated part of the new overall approach to design. The overarching objective of this new approach is assuring that resources necessary for production will not be diminished over time. This is really just a statement of the conservation of matter and energy, but applied to a much broader circle or sphere of influence than the inside battery limits (ISBL) of a CPI plant to include perhaps the entire planet! (Of course, this assumes our friend the Sun will be around for a few more millennia providing free energy.)

The overall objective of this book is to introduce SE as the logical next phase of continuous improvement to process and product engineering design for manufacturing in the chemical process industry (CPI). New product design will lead the way to SE through the utilization of sustainable materials, while an existing, robust process design methodology will assure that material and energy optimal designs are implemented. The two go hand in glove.

An already fairly complete chemical process design literature exists from which to move forward. This literature describes a classic engineering design methodology, at once recognizable and largely successful up till now, throughout the CPI including petrochemical refining. The CPI is a net positive exporter of US products, no better record of success can be had, so changes should be made only as needed. Systemic improvement to classic design has levelled off, and now SE is poised to propel the industry into a future that combines permanence, profitability as well as livability.

The seeds for SE were planted in 1970 when EPA and OSHA regulations came into being, and we have been on a track of continuously improved, cleaner, safer, and profitable CPI process designs ever since. Along the way, it became apparent that often on the road to meeting those environment, safety, and occupational health (ESOH) regulations, generally through process equipment or operations modification, bottom-line performance was also improved. It seems simply questioning why something was being done the same way for the past 40–50 years led to process improvements and attendant cost savings. As the 3-M company has said "Pollution Prevention Pays" copyright 3M.

The Congress entitled the primary chemical manufacturing regulation, the Resource Conservation and Recovery Act (RCRA). So this might be headwaters of sustainability engineering, as it covered not only materials but energy as well—just read it!

Pollution Prevention (P2) on the part of the CPI has led to a much more efficient use of resources at their point of use, the manufacturing plant. But just as increasing the miles per gallon (MPG) of automobiles from 10 to 15 in the 1960s to perhaps 50 within the next 10 years won't increase the supply of nonrenewable gasoline, something else is needed to get to the next level.

More recently, an additional positive unexpected consequence of CPI P2 and ESOH regulatory compliance has arisen, namely the public demand for all things Green. Everything from more efficient refrigerators and automobiles to food grown naturally, in short, everything consumers can lay their hands on now has a Green focus.

And the concept of all things Green is now morphing into all things Sustainable. The demand for consumer products with an environmentally neutral production footprint is high. Whereas environmentally neutral once simply meant not polluting, it now means having little to no effect on the surroundings, hence sustainability, or, leaving things pretty much they were prior to manufacturing.

Product labelling such as EPA Energy Star and food nutritional labels have created educated consumers willing to shop and pay for improved energy savings and personal health. In turn, this has created a consumer market for things that are responsibly made and work better. In the past, there was no way for consumers to evaluate such claims without trial and error. Now a wide variety of fact-based information is available that doesn't require a degree in Science or Engineering. The only endangered species here are the "Snake Oil Salesmen."

Engineers in the CPI have always been charged with designing the best, safest, and cheapest manufacturing processes. Process is at the heart of all chemical engineering, and since 1970, both process and process design have been improved to incorporate EPA and OSHA up front. Chemical engineers "own process," and the entire CPI workforce is in a positon to bring the sustainability ethic and a return of pride of workmanship or ownership which also leads to product and economic improvement on the process side.

The conservation laws governing matter and energy, fundamental to science and engineering alike, along with a dose of thermodynamics, constantly remind us that 100% conversion of anything is impossible. (Good luck to those still looking for the perpetual motion machine!) But SE is really just another process design parameter improvement to push back the event horizon of diminishing resources. Life Cycle Analysis (LCA), a P2-inspired improvement to classic engineering economic analysis, was suggested by EPA in the early 1990s. LCA broadened the design review from the immediate process to include a cradle-to-grave accounting of inputs and outputs.

Action to integrate all past advances with key SE supporting elements to classic design includes:

1. Integrated power generation and chemical production in order to squeeze the last drop of both high and low-quality energy available in material streams, a feat best accomplished in the CPI, for small as well as large-scale industry.
2. Renewable material and energy resources utilization whenever and to the greatest extent possible, including incorporation of *all* waste materials into useful products or recovered energy, or feedstocks elsewhere, even offsite.
3. Efficiency improvement in all things used to manufacture, such as motors, pumps, turbines, and heat exchangers, will be required. Tighter process control between disparate but linked processes will demand less variability in efficiency with load.
4. Quality management—more important than ever with this high level of interconnectedness.
5. SE trained and educated workforce focused on sophisticated, optimal solutions first, not later.
6. ESOH Compliance.

To assure the success of SE into the future, and in keeping with ABET (academic accrediting board) recommended use of design practitioner-instructors, a practice-based approach to SE within the university academic design curriculum is badly needed. Accordingly, I have included a few SE-oriented examples from the design course I taught as an adjunct professor for 6 years at the University of Illinois at Chicago. I drew upon my own work experience as well my service with the National Council of Examiners for Engineering and Surveying (NCEES) to identify the Chemical Engineering PE exam specification as a reference for teaching practice-based design as presented herein. It should be noted also that new product design, distinct from process design, has a home in academia just as in industry R&D labs. Process and product design are ultimately linked together, and it is hoped my academic colleagues as well as CPI industrial trainers might also find this work of some use.

Also important is the need for a highly trained and skilled technical workforce. The present shortage of such skilled individuals has hurt US manufacturing. With such a high degree of automation today, there is little need to export as much manufacturing as we already do, so much can come back. Skilled workers, along with engineers, will be needed to accommodate this. Sadly, the high school shop courses I took that once served as a springboard to skilled jobs and that also helped shape and inform my own engineering skills have all but disappeared.

A section on SE in the other engineering disciplines, e.g., civil, structural, mechanical, electrical, and environmental, is also included. In the Engineering, Procurement and Construction Industry, the disciplines actually carry the major portion of the total project budget. And within non-CPI industry segments, much improvement toward SE is coming from enhanced road building and construction techniques, more efficient HVAC, motors and automobiles, and lighting and power electronics and increased incorporation of recycling as an energy and materials feedstock.

Overall Book Summary: How It All Fits Together

Chapter 1 will develop some background and historical framework for SE.

In Chap. 2, we look at SE and how it works within the existing process and product design framework.

Chapter 3 examines a key principle of SE, i.e., the use of renewable resources, both energy and materials to assure, well, sustainability!

All design engineers in general are quite familiar with equipment efficiency, but SE, with its higher degree of coupled process and power integration, demands a much closer evaluation of the Efficiency of Everything found in Chap. 4.

Chapter 5 provides an overview of the EPA pre-manufacturing notice (PMN) requirements. Prior to embarking on any new product design, it must be examined to both assure permitability and identify potential, less toxic, and cheaper alternative methods and chemicals, and the PMN process is the place to start.

Chapter 6 provides a cursory overview of the principle regulations pertaining to the manufacture of chemical products in the United States.

Chapters 7 and 8 look at basic SE tools and process examples respectively to give an initial point of departure on how to apply SE basics.

In order to optimize all engineering endeavors to the positive, the entire "affected community" must be involved. Design engineers, client engineers and management, regulators, and the community into which the plant will operate are all included. For this reason I have added Chap. 9 on Quality Management.

The federal and state MPG and Energy Star requirements have been enormously successful over the past 40 years in saving energy and consumer dollars. By creating a level playing ground, industry responded by creating much more efficient cars and appliances. Despite the early doomsayers, things are now cheaper to make and operate, as well as more profitable for industry that sells them. When all is said and done, there may or may not be a need for new regulations and/or standards to promote SE, similar to these above, and Chap. 10 will get that dialog going.

Although the book is aimed primarily at the chemical process industry (CPI), there are numerous SE examples from other engineering disciplines and industry segments. Chapter 11 takes a cursory look at some of these with an eye to encouraging others to follow, as well as a reminder to CPI types of the need to cooperate closely with all disciplines to come up with the best SE designs.

Chapter 12 provides a summary and overall resolution roadmap to SE. In addition to a recap of the previous 11 chapters, we take a look here at three non-design areas in need of continuous improvement updates focused around training and education, namely:

1. A return to the preparation of a technically competent manufacturing workforce
2. Focus on minimum industry practice competency standards
3. Fine-tune baccalaureate academic preparation, harmonized with industry standards

In the USA, it seems, we have all but given up on supporting manufacturing and rather lean toward a self-fulfilling prophecy of making everything offshore. If all of the affected community does not have a seat at the SE design table, the outcome will be suboptimal at best. One of these three missing links to SE is the lack of a skilled workforce capable of operating, maintaining, and, by their very presence at the table, improving manufacturing operations, thereby reducing costs and altering the "where to make it" equation. Appendix A provides some reference material on the subject.

Finally, IMHO, industry, and academia have been drifting apart, at least in the method and manner of preparation of our engineering workforce. The academic emphasis has been more on the theoretical, and this shows up in the course content, at both the baccalaureate and graduate level. While this research orientation bodes well for new product and process development, if SE design-based manufacturing is to take hold, it will depend in the long run on adequate training and education of our young men and women in a more practice-based manner (even ABET has recognized this regarding design education!).

A more detailed review of these two missing links is presented in Appendix B and C, with a review of the NCEEES PE Exam process, which I believe represents the industry "Standard" as well as an example of an SE-based senior design chemical engineering curriculum from the two-semester course I taught as adjunct professor at the University of Illinois at Chicago for 6 years. Hopefully this constructive criticism will encourage industry to engage more closely with local universities and take a closer hand in developing the talent of the future.

Ultimately SE will be a boon for industry and consumer alike, just as those brought about by the ESOH changes of the 1970s. As the entire world is now going into the production and consuming game, greater demand is being made on fewer resources. SE is just a way of improving our systems of the past to ensure we have a future. And the past has shown this can be done safely as well as profitably.

Lincolnwood, IL, USA
2024

Jeffery P. Perl

Acknowledgments, First Edition

Just a few thanks are in order here, but they are important.

Much of what is presented herein comes from 40 odd years of work, ruminations, and many conversations about things of importance to society at large, from an engineers' perspective of course! I talk a lot and learned much and hopefully gave something in return … dialog, not monolog is the key here!

From Springer—Tiffany Gasbarrini for bringing the title "Sustainability Engineering" and for hiring me to fill in the blanks, and Zoe Kennedy, her assistant, for baby-sitting me throughout. The Springer Production Team, Sharmila Kirouchenadassou, and her boss Dhanuj Nair did a great job turning my combination of typing and scribbling into the readable text you see here. Any mistakes are purely mine.

Thanks to Dennis O'Brien, PE, "first amongst equals," for discussions regarding: the need for engineers to plan for global warming, regardless of the cause, e.g., coastal degradation from rising sea levels and poorer cooling tower performance, to name a few, that led to the notion of Enlightened Self Interest, and for editing portions of the text. My friend Dan Rusinak, PE provided a wealth of chemical engineering resource material over the 6 years as he served as "Imagination Wizard" for the ChE senior capstone design course I taught for from 2008 to 2014 at the University of Illinois at Chicago. A dozen or so of my Chicago area design professional colleagues helped out with the course are listed in the appendices. They set a perfect example of a healthy, much needed, industry-academia interface. Their participation was another example of enlightened self-interest as many of my students were ultimately hired by their companies!

My chemical engineering education at the Illinois Institute of Technology provided a strong basis for going forward in so many ways. While a graduate student, I studied "Energy and Society" under the late Prof. Henry Linden, past president of IIT and Gas Technology Institute, bringing in his friends such as Edward Teller and Amory Lovins to talk about nuclear and soft energy, respectively. A special appreciation to Prof. Darsh T. Wasan, my doctoral advisor and research guru, for providing the opportunity and exposure to such a broad breadth and depth of subjects extending well beyond the pure engineering.

As usual, all errors are my own, and I trust the readers will provide constructive feedback.

A final thanks to my family and office mates who put up with my rantings, ravings and mutterings over the past 2 years as I agonized over what to put in and what to leave out.

Chicago, IL
2016

Acknowledgments, Second Edition

My sister Karen Perl gets the main award this time for both encouraging her big brother to maintain his library after closing his Chicago office this past year and for insisting he accept the offer from Springer to update the first edition, my Brother/Lawyer Allen for watching over the contract, and for my family putting up with sound blackouts while I wrote this past year.

A big thank you to Michael McCabe, Executive Editor, Applied Sciences at Springer for suggesting and guiding this effort and for his team at Springer NY; Brian Halm as well as Bakiyalakshmi RM (Ms), for Springer Nature, Straive, Production Editor (Books) for editorial advice and requests and Sumathy Thanigaivelu. But wait, there's more... I just completed review of proofs so any mistakes are my own, but must recognize Kala Palanisamy, Straive Production Supervisor (Books) and my project manager at Straive, and her team for calling attention to numerous things for me to address (took me 10 days!) that will, I am sure, make this work more accurate and readable.

Lincolnwood, IL, USA
2024

Jeffery P. Perl

Contents

Abbreviations

A&E	Architects and Engineers
ABET	Accreditation Board for Engineering and Technology
ACS	American Chemical Society
ADM	Archer Daniels Midland
AFDC	Alternative Fuels Data Center
AFUE	Annual fuel utilization efficiency
AgE	Agricultural engineering
AIChE	American Institute of Chemical Engineers
AKA	Also known as
ALPHA	Distillation separation factor
ASHRAE	American Society of Heating, Refrigeration and Air Conditioning Engineers
ASME	American Society of Mechanical Engineers
ATR	Auto thermal reformer
BEC	Base Environmental Coordinator
BFD	Block flow diagram
BiomedE	Biomedical engineering
BOE	Barrel of oil equivalent
BOF	Basic oxygen furnace
BPSD	Barrels per stream-day
BRAC	Base Realignment and Closure Act
Btu	British thermal unit
Btu/h	Btu per hour
CAA	Clean Air Act
CAD	Computer-aided design
CapEx	Capital expenditure(s)
CAS	Chemical Abstracts Services of the ACS
CBI	Confidential business information
CE	Civil engineering
CEP	Chemical engineering progress
CESQG	Conditionally exempt small quantity generators

CFL	Compact fluorescent light
CH_4	Methane
ChE and BioE	Chemical and biological engineering
CHP	Combined heat and power
CO	Carbon monoxide
CO_2	Carbon dioxide
COE	College of Engineering
COP	Coefficient of performance
COTS	Commercial off the shelf technology
C_P	Heat capacity, constant pressure
CPI	Chemical process industry
CWA	Clean Water Act
CY	Calendar year
Delta T	Temperature difference
DES	Deep eutectic solvents
DOE	Department of Energy
DOT	Department of Transportation
DRC	Democratic Republic of Congo
DRI	Direct reduction of iron
DVD	Digital video disk
E&C	Engineering and construction
ECM	Electronically commutated motor
ECOSAR	Ecological structure activity relationships program
EE	Electrical and electronics engineering
EER	Energy efficiency ratio
EERE	DOE Office of Energy Efficiency and Renewable Energy
EG	Ethylene glycol
EIA–US	Energy US Information Agency
EIT	Engineer in training
EnvE	Environmental engineering
PEO	Presidential Executive Order
EPA	Environmental Protection Act/Agency
EPC	Engineering, procurement and construction
EPCRA	Environmental Protection and Community Right to Know Act
EPISuite	Legacy EPA program containing ECOSAR
ESIH	Environment, Safety and Industrial Hygiene
ESOH	Environment, Safety and Occupational Health
ESOH-CAMP	ESOH Compliance and Management Practice Program, USAF
EtOH	Ethyl alcohol
EWB	Engineers without borders
FAA	Federal Aviation Agency
FDA	Food and Drug Administration
FDR	President Franklin D Roosevelt
FE	Fundamentals of Engineering Exam
Fe	Iron

Fe_2O_3	Iron oxide
FTL	Fischer–Tropsch liquids
GMO	Genetically modified organism
GS	Ground Source
GW	Gigawatts
H_2	Hydrogen
H_2N- $COONH_4$	Urea
HBI	Hot briquette iron
HEN	Heat exchanger network
HETP	Height equivalent to a theoretical plate
HHS	Department of Health and Human Services
HID	High-intensity discharge bulb
HNO_3	Nitric acid
HP, hp	Horsepower
HQ	Headquarters
HRSG	Heat recovery steam generator
HVAC	Heating ventilating and air conditioning
I&EC	Industrial and Engineering Chemistry Division (ACS)
ICE	Internal Combustion Engine
IDHL	Immediately dangerous to life and health
IE	Industrial Engineering
IEDO	Industrial Efficiency and Decarbonization Office of USDOE
IFC	Issue for Construction (Drawings)
IGCC	Integrated gasification combined cycle
IIT	Illinois Institute of Technology
IMHO	In my humble opinion
IRP	Installation restoration program
IRR	Internal rate of return
ISBL	Inside battery limits
ITP	Doe Industrial Technologies Program
JCAP	Joint Center for Artificial Photosynthesis
Kw	Kilowatt
LC	Level control
LCA	Life cycle analysis
LED	Light emitting diode
LHV	Lower heating value
LMTD	Log mean temperature difference
LNG	Liquefied natural gas
LDES	Long-duration energy storage
LOTO	Lockout/Tagout
LOUC	Law of Unintended Consequences
LRA	Local Redevelopment Authority
M	Mass
M&E	Material and energy
MatBal	Material balance

MBPE	Material balance and process effluent plant
ME	Mechanical engineering
MetE	Metallurgical engineering
MEV	Multiple effect evaporation
MIT	Massachusetts Institute of Technology
MMAE	Mechanical, materials and aerospace engineering
MMSCFD	Million standard cubic feet per day
MOC	Management of Change (OSHA PSM)
MPG	Miles per gallon
MPGe	Miles per gallon – electric equivalent
MSW	Municipal solid waste
MVR	Mechanical vapor recompression
N2	Nitrogen
NAFTA	North American Free Trade Agreement
NATO	North Atlantic Treaty Organization
NBS	National Bureau of Standards
NCEES	National Council of Examiners of Engineers and Surveyors
ND	North Dakota
NEC	National Electrical Code
NFPA	National Fire Protection Association
NG	Natural gas
NGL	Natural gas liquids
NH3	Ammonia
NH_4NO_3	Ammonium nitrate
NIMBY	Not in my backyard
NiMH	Nickel metal hydride battery
NIST	National Institute of Standards and Technology
NPV	Net present value
NREL	National Renewable Energy Lab, DOE
NWU	Northwestern University
O&G	Oil and grease
O&G	Oil and gas (companies)
O&M	Operations and Maintenance
O2	Oxygen
OPPT	USEPA Office of Pollution Prevention and Toxics
OSBL	Outside the battery limits
OSHA	Occupational Health and Safety Act/Administration
P&ID	Piping and instrumentation drawings
P2	Pollution prevention
PA	AF public affair
PC	Personal computer
PCB	Polychlorinated biphenyls
PDF	Portable document format
PE	Professional engineer
PEL	Permissible exposure limits

PF	Power factor
PFAS	Per- and polyfluorinated substances
PFD	Process flow diagram
PG	Propylene glycol
PMN	Pre-manufacturing notice
PNNL	Pacific Northwest National Lab
POTW	Publically owned treatment works
PPA	P2 Act
PPM	Parts per million
PSC	Permanent split capacitor motor
PSM	Process safety management
PV	Solar photovoltaic
Q	Heat
Q&A	Question and answer
QM	Quality management
R&D	Research and development
RAB	Restoration Advisory Board
RCRA	Resource Conservation and Recovery Act
RDF	Refuse derived fuel
REACH	Registration, Evaluation, Authorization and Restriction of Chemicals Program (Europe)
ROD	Record of decision
ROI	Return on investment
RPM	Revolutions per minute
STEM	Science, technology, engineering, and math
SAR	Structure-activity relationships
SBIR	Small business research and innovation funding
SE	Sustainability engineering
SEATO	South East Asia Treaty Organization
SEC	US Securities and Exchange Commission
SEER	Seasonal adjusted EER
SFP	Specialty Fertilizer Products, Inc.
SGTL	Syngas to liquids
SMM	Sustainable materials management
StE	Structural engineering
SUO	Swing unit operations
TC	Toxicity characteristic
TDH	Total developed head
TIC	Total installed cost
TPD	Tons per day
TPP	Trans Pacific Partnership
TQM	Total quality management
TSCA	Toxic Substances Control Act
TSDF	Treatment Storage and Disposal Facility
UF	Ultrafiltration

UHV	Upper heating value
UIC	University of Illinois at Chicago
UK	United Kingdom
UL	Underwriters Laboratory
UN	United Nations
Uoa	Overall Heat Transfer Coefficient
USAF	United States Air Force
USDA	United States Department of Agriculture
USEPA	United States EPA
VCR	Video cassette recorder
VFD	Variable frequency drive control
VOM	Volatile organic material
WTE	Waste-to-energy
WWII	World War 2
WWTP	Waste water treatment plant
XP	Explosion proof

Chapter 1
Introduction: Enlightened Self-Interest for the Enthusiastic Capitalist

1.1 Sustainability: The New Process Engineering Design Optimization Parameter

In the 1970s, the chemical process industry (CPI) began to incorporate the newly minted Environmental Protection Agency (EPA) and Occupational Safety and Health Administration (OSHA) regulations into process design for new and existing manufacturing facilities. That was the first major tune-up since the inception of chemical engineering (ChE) in the late 1800s. Sustainability engineering (SE), the topic of this book, provides the next "Continuous Improvement Tune-Up."

1.2 Punch Line: For All My Fellow Engineering Colleagues – Please Take Heart

The tried and true design methodology used to build and maintain the chemical processing industry is alive and well! You are not being sent off to the Gulag for retraining! Consider sustainability engineering as a continuous improvement module. Finding ways to use more renewable resources, both energy and materials, is the bottom line. Also, it does not matter which side of the global warming issue you are on. The fact is, as temperatures are increasing and sea levels are rising, we engineers are called upon to include these factors in our designs, particularly in coastal regions or wherever temperature matters. We can ignore the blame game and simply design for new ambient conditions that impact heat exchange, flood considerations, etc., something we have always done anyway. By definition, SE design and manufacturing reduces CO_2 footprint as well as natural resource consumption use while saving money as well. Stick to fact-based decision-making (USAF Surgeon General, ca. 1998, "Fact Based Decision Making," Private communication) and park your emotions at the door.

J. P. Perl, *Sustainability Engineering for Enhanced Process Design and Manufacturing Profitability*, https://doi.org/10.1007/978-3-031-52363-2_1

1.3 A Bridge to Tomorrow

As a chemical engineer, I have had a lifelong interest in responsible manufacturing. I came of age in the wake of enormous man-made environmental catastrophes such as Love Canal, Hudson River PCB dumping, Waukegan Harbor and the infamous "Valley of the Drums" that helped lead to the EPA Superfund regulation. I guess my favorite was the Cuyahoga River in Ohio setting on fire. Around that time, Rachel Carson's book "Silent Spring" was published [1]. Her testimony to Congress (ca 1962) along with the incredible environmental catastrophes such as the Cuyahoga River catching on fire got so bad that Republican President Nixon jumped on the Earth Day 1970 bandwagon and signed into law the creation of USEPA to protect the public environment and US OSHA to protect the workforce, both in 1970. A government colleague, whom I contracted to for a year while working on his report to congress re: Superfund [2], and I presented at a 1990 Milwaukee environmental conference he and I were attending. There, he announced the biggest threat to pollution prevention was "chemical engineers over the age of forty." I thanked him for grandfathering me in as I was only 39 at the time!

He was a purposely incendiary speaker, who was engaged, at that time, in his role at the US Office of Technology Assessment (OTA), advising the Congress on creating pollution prevention regulations. I had to agree with him, because from my observations at the time, the existing CPI design process often showed waste by-products on the piping and instrumentation drawings (P&IDs) going to "other" on the drawings … trouble was, there wasn't always anything connected to other, and little care was given beyond product creation! So, waste often went out the back door or into the "back forty" allowing short-term profiteers to benefit until modern EPA regulations with teeth came into play.

Fast forward to today, and of course compliance with EPA and OSHA are now second nature, but an unintended consequence is a legacy of aging laws that are occasionally at odds with present modern principles of sustainability engineering presented here. These laws will need to be realigned to provide industry and communities greater flexibility toward adopting sustainability engineering design principles, while maintaining and improving public health and safety as originally intended. Per USEPA [3] Resource Recovery and Conservation Act (RCRA) regulation, a material is not a waste until the generator declares it thus, so why not continue to use it either onsite or elsewhere. As we see with SE, there is little to no reason for anything to go unused. All chemical manufacturing has the ability to either turn raw material into finished product, or recycle into energy or convert to another merchant product, e.g., hydrogen for use elsewhere, with little to no regulatory development. Fine-tuning might be needed.

In 1990, Congress was in the midst of trying to reauthorize some of the USEPA regulations that included, among others, a definition of pollution prevention, an area that CPI had done poorly up to that time. I argued then and still believe that pollution prevention (P2) is a continuous improvement activity that could bring a net positive value to the manufacturing process if only a proper boundary was drawn in space and time around not only the process but also the affected community. It was also important to allow industry to recycle/reuse its own waste without the unnecessary hindrance of Treatment, Storage and Disposal Facility (TSDF) regulations. I came to this view from total quality management (TQM) principles [4].

A bit later, Motorola won the Illinois Governor's Pollution Prevention Award while finding a new, better, and cheaper way to solder electronic boards in 1994 and asked me to help them understand what they were doing right [5]. I found they attacked P2 as they would any production problem by treating pollution as a defect. A corporate policy entitled "Going Beyond Compliance" was the program my colleagues at Motorola adopted, in order to meet this new demand. Motorola's chief success was based on its quality orientation that involves contribution from everyone at all levels of the company in this program, and this is what is necessary for SE to succeed as well.

For all those who have looked at the chemical and related manufacturing industry and feel there are better, safer, more profitable ways to do things, this book outlines a uniform approach to SE. With the creation of EPA and OSHA, the CPI adjusted to a level playing ground that provides greater protection to worker and consumer alike. Over the past 10-15 years or so, the Industry was caught quite off guard, pleasantly, by citizen interest in all things "Green." But while the Industry has done a pretty good job with compliance, it turns out that simply doing all the above is not quite enough. Necessary, but not sufficient, as my math buddies say.

Enter "sustainability," a new buzzword that has a positive meaning but perhaps not always so well defined. This too revolves around TQM principle of the quality team and continuous improvement. Running out of materials and energy can be viewed as an overall defect, so long as the issue is not pushed "off the P&ID" to others for consideration. It turns out that the only "other" left is us. As Pogo, a cartoon character of the pundits of the 1970s famously said, "We have met the enemy and it is us" (Walt Kelly, Cartoonist, Pogo).

While still in school, one of my professor mentors told me to stop working on a problem so hard if a solution was not at hand. He taught me the importance of putting it down for a while, working on something else, and then coming back to it. I turned that notion into a metaphor, viz., "When scaling a mountain, the team often will move in fits and starts, upward always until progress is halted by some unknown, unexpected challenge, and a new basecamp is established, often requiring lateral moves until an upward direction can once again be resumed." (Jeff Perl) The key here is to never lose sight of the end goal or objective, and always to adjust as necessary.

Now, it appears that the earth's material and energy resources may be finite. Our understanding of the biosphere within which we live has been greatly sharpened over the last 40–50 years. In addition, our populations continue to grow with ever-increasing load and demand on planet earth for simple sustenance. So, now we must draw our quality circle around a non-process entity, i.e., the earth, as a member of the affected community that asks the question … is it sustainable. Fortunately, engineers can factor this into bottom-line economics regardless of the cause of climate change. Sustainability engineering meets long-term economic performance requirements, by including the new variables as they present themselves. Sustainability implies permanence, but forever is a long time and the definition of infinity here may be more like 50–100 years—enough to bridge us to the next major improvement. And just because we do not know what that will be is no excuse to avoid aiming for it.

For many years now, the chemical process industry (CPI) has recognized the need to assure a supply of raw materials prior to constructing vastly expensive petroleum refineries for example. But most of these resources by definition are not renewable. At the very least, a manufacturing process is not internally sustainable if there are insufficient raw materials and energy available to it over its project life. So the quality circle now loops around a rainforest in the Amazon, or metal and gem mines in South Africa. It also loops around the Environment, Safety, and Occupational Health (ESOH) considerations of the heretofore unexamined community involved with providing the raw materials in far-off places, not controlled by our regulations. Regardless of political or philosophical concerns, the degradation of one community to serve another simply leads to long-term acyclic disruption in a globally interactive world. (Probably more the purview of governments, and is most emphatically not sustainable in the long term.)

In the USA, we came to realize that pre-regulatory CPI management practices were not sustainable. Los Angeles in the 1970s produced teary eyes as well as impacted lung capacity, and that was just mostly from automobile exhaust! London's coal burning once aggravated, if not occasionally caused, its famous fogs. China now also has begun to realize they cannot live in a totally unregulated industrial manufacturing environment and, in some instances, are forced to import safe foods. As I write this, China and the USA signed an agreement regarding carbon trading that has been useful by turning CO_2 into a stock exchange commodity. China is also learning that spill and accident prevention and air and water emissions will need to be controlled. Here, in the USA, civil and criminal fines and penalties were required for ESOH compliance and what form this takes in China will be interesting to see. Our methods may be different, but in the end we all need to eat, drink, and live in a safe environment, and in a world of finite resources, to do it sustainably. The new Trans-Pacific Partnership (TPP) and its international trade policies will encourage and help developing countries to sustainably husband their resources.

A 2023 update: The prior administration took the USA out of the TPP. Predictably, the other member nations went ahead with this on their own.

At COP26, the USA and China issued a joint declaration committing to cooperate on a variety of climate change issues, including developing regulatory frameworks, and environmental standards to reduce greenhouse gases emissions, encouraging decarbonization and electrification of end-use sectors, and fostering the circular economy (2021).

It is challenging to keep up with international agreements, but one thing is clear, a sustainable world will require some reasonable degree of world cooperation.

Within the US borders, sustainability engineering ties this together in a new process design optimization paradigm that can inform capitalism to better and more profitably serve itself. As a side note, USEPA and OSHA regulations already consider long-term economic impact when setting regulations over the last 45–50 years. Process design engineers need to be aware of these additional economic design parameters or risk mid- to long-term project failure for the companies that rely on their Net Present Value (NPV) project go/no-go evaluations. All can engage in this new way of going forward. And in the greater scheme of things, this is a relatively minor tweak to the way the CPI operates today compared to the huge change brought by the ESOH regulations of the 1970s. SE just requires greater detail to engineering and estimating everything which can improve economics.

1.4 What Is Sustainability Engineering All About?

In the USA, the imposition of ESOH regulations on the CPI 50 years ago, equally across the board through federal regulations, has stood the test of time quite well. A resulting focus on pollution prevention (P2) and quality manufacturing has actually made our products cheaper and much more reliable. We now have a scientific fact–based methodology for setting safe air, water, and ground contamination levels. The workplace is now also much safer as are surrounding communities. The existing ESOH laws protect all Americans, rich and poor alike, and federalization has made it possible to spread these costs equitably across the board.

But we have reached a sort of point of diminishing returns for the system as it is now. We are still climbing the mountain, establishing a new base camp as we focus on a sustainable safe workable fix, and continue upward. I believe we are at this point now as it relates to the method and manner that our socio-industrial system can support its population base. So what is next?

The existing design methodology for the CPI works quite well overall. SE is merely the next step in a continuous improvement process. The challenge is simple: maintain a balance on the planet that assures what the ChE calls steady state in perpetuity. We do not have a perpetual motion machine, so we need to work with what is all around us. The sun is here and, thankfully, photosynthesis. Our knowledge of thermodynamic machines, sustainable crops, etc., will also help.

As President Harry S. Truman said, "There is nothing new in the world except the history you do not know" [6], and it is certainly true for SE. In the seventeenth century, Galileo was excommunicated for his science-based view that the earth revolved around the sun. Eventually, the weight of science finally fell to his theory. But it is worth noting that all the excellent work in physics, math, and astronomy that came before Galileo was mostly all still useful. Fortunately, the wheel did not need to be reinvented, and the new focus on the heliocentric perspective led to a blossoming of "fact-based" science free of superstition. The French Enlightenment followed and set us on the course we are on now, including the founding of our country. We will try to stick to the facts throughout this book as that seems to work well!

1.5 Guiding Principles of Sustainability Engineering: From Present to Future

And as President Truman said above, the same is true as we move towards an SE future. Not much really new is required, all the hard won, existing, basic, and sophisticated engineering design principles still apply. So, what has or will need to change?

Once the primary design requirements have been delineated, an SE decision process will use at its core, classic process design components. The SE designer will need a renewed, refocused, and deeper reliance on the following design elements.

1.6 Enter Basic Sustainability Engineering Design Elements

1. Use of *renewable material* and *energy resources*:
   (a) Hey, your plant would not run without these anyway.
   (b) Quick scrub to assure no ESOH issues and that alternate, safer chemistry cannot do the job.
2. Seek out *combined power* and *chemical production* opportunities to maximize *resource conservation*:
   (a) Save on local power generation and energy reuse.
   (b) Use materials and energy in between disparate processes in one location.
   (c) Minimize waste through recycling—The Undiscovered Feedstock!
3. Enhanced focus on *efficiency improvements* or risk suboptimal SE design:
   (a) Whether or not global warming is a large percentage of temperature rise is not the point.
   (b) Science seems to support this, but it does not matter; resources and production are still adversely affected by it, whatever the cause.
   (c) More efficient processes and emphasis on renewables will make this debate a moot point.
   (d) The bottom line must be maintained for the US industry to continue to thrive and grow.
4. Drill down deeper on ESOH to ensure *minimized effect on the environment*:
   (a) This will require extending the calculation boundaries way beyond inside battery limits (ISBL).
   (b) Use of the EPA developed Ecological Structure Activity Relationships (ECOSAR) model, for creating better, safer, cheaper manufactured chemicals.
   (c) Aim for "Zero Discharge"; it can actually save money in the long run.

5. *Education improvements* needed for SE:
   (a) *Existing* education and training is insufficient and *inadequate* for SE.
   (b) Adopt more *practical* production versus *research* baccalaureate tracks. University research now dominates academia and new engineers are less prepared for industry or SE.
   (c) Return to *trade school* and community college model to create workforce organized around SE:
      - Heat pumps, photovoltaics, solar power, and mini-power generation all will *need highly skilled technicians*, more so with complex SE interacting systems.
6. *Capitalism refined and improved.* Corporate profit-sharing based on longer-term goals will improve, not reduce, profits. Quarterly report system needs updating or SE is doomed from the start. Deming taught this to the Japanese with a *focus on quality* and people (Chap. 9, Sect. 9.8).
7. All successful businesses use some form of *quality management* (QM).
   (a) Good business requires it, SE demands it!
   (b) SE complexity is manageable but needs to incorporate total quality management (TQM), or equivalent, principles to assure success.
   (c) Driver—Sales and profit rewards are there as the public demands all things green.
   (d) Industry saved much money by preventing pollution just as 3M moto says:
      - "Pollution Prevention Pays."
8. As *SE manufacturing* becomes more *efficient* and *self-sustaining*, it *creates new business* lines:
   (a) This leads to *increase* in *domestic manufacturing* operation opportunities.
9. Now is the time to move toward *SE-centric systems* as we continue to climb the mountain.

Over the last 10 or so years, the concept of sustainability has crept into the public consciousness. In a general sense, the definition refers to something that can, in some way, replenish itself. This includes not only the materials but also the energy needed to sustain production, and, in turn needed to sustain life as we know it on planet earth. That means forever, not just the typical 2- to 3-year payback desired by companies, or the 10- to 20-year life expectancy of new CPI manufacturing plants. Now, the applied mathematicians would argue that infinity is a matter of definition, and perhaps 100 years might be long enough. But that is a far cry from what we have now and SE design will require, well, sharpening our pencils and computer modelling programs a bit more. I believe we can do this and the challenge is a perfect new task for government-funded academic research and training programs (beyond basic science).

Agriculture, as an example, appears on the surface to be sustainable, i.e., farmers grow and harvest crops every year in a seemingly endless manner. But the level of effort including, but not limited to, pesticide application, GMO modifications, soil erosion prevention, and water conservation is not necessarily sustainable. When viewed in the greater perspective of a growing population, the cost of all these activities and others, e.g., antibiotic-resistant farm animal diseases are not sustainable in the medium- or long-term. SE design requires that a larger circle be drawn around the affected community in order to illuminate what can and should be done as well as obtain input on how best to do it. Some government assistance to small farms as well as R&D for all will pay great dividends. Agriculture will benefit greatly from SE practices, technically as well as financially.

1.7 The Quality Circle Approach

In our own way, process design engineers have been doing sustainability engineering ever since the discipline of chemical engineering (ChE) came into being some 140 years ago. In those early days, the designer was simply concerned with the economic success of a particular manufacturing facility. The boundary of this facility was the fence line and the economic movement of raw materials into and finished products out of it was the primary concern. The process efficiency was probably an early concern as raw materials and energy, although abundant, were certainly not free. However, little to no concern over environment and worker safety was in evidence beyond simple, obvious ones, such as preventing costly explosions (DuPont managers lived onsite of explosives factory! ca 1806, possibly the world's first process safety management program!) and exposure to chemicals immediately dangerous to life and health (IDHL), both which visibly led to huge short-term expense.

Roughly beginning in the 1880s, the new discipline of chemical engineering quickly set up early design rules governing material and energy balances, which, with the nascent field of "unit operations" could be reduced to unified, science-based engineering design equations. These rules began to codify and mathematically describe the chemical manufacturing process and led to the early identification of energy as the equal partner to materials in design. To me, the material and energy balance concept is a sort of double-entry accounting system if you will, where material conversion and flow could be related to the necessary energy consumed or produced (conservation of mass and energy).

Early capitalism was unable to cope with the then emerging "modern" industrialized world with protection clearly lagging behind profits. Strip mining and solution mining left large tracts of land forever unusable. With no economic penalty for corporate environmental behavior, they did the best they could/should. The advent of flight allowed us to see the destruction in perspective. (An unexpected positive consequence from Orville and Wilbur and early aerial photography [Sanborn maps] were used by insurance companies). This early industrial age short-sighted approach was not much different than the past agricultural methods of civilizations that overused and rendered land useless while creating deserts … all preventable if a big picture had been taken. It is also a great example of conversion of an unknown unknown into a known unknown. We now know what we did not know before and must accommodate and incorporate that understanding. (Chap. 9, Sect. 9.4)

Back then, little to no attempt was made either to prevent pollution, or husband natural resources as unregulated waste disposal out the back door, into rivers and ground and air, was essentially free while raw materials and energy supplies were abundant and seemingly limitless. Indeed, entire primordial forests were clear-cut, coal and other minerals mined from the earth with methods of total disregard for modern environmental, let alone sustainability concerns. Actually, the word "disregard" may be too strong, because taken within the economic and regulatory context of the early days, 1800s, there was little economic or regulatory reason to be concerned. The connection between occupational health and disease that would eventually lead to ESOH regulations was just beginning to be uncovered and explored, e.g., coal miner's lung, mercury vapor poisoning of the central nervous system (the Mad Hatters disease), where people painting with Radium to create glow in the dark watch dials were dying from radiation induced cancer. Pasteur had only recently made the connection between diseases caused by bacteria. Mid-nineteenth-century English Physician John Snow correlated disease outbreaks, e.g., typhus with the proximity of human and animal waste to drinking water supplies [7]. Clearly, those urban living models were unsustainable as population density grew without regard to what is now considered basic hygiene practice. The connection between industrial air pollution and lung disease and chemical water pollution and other bodily insults would take considerably longer to make; as would the need to protect the vegetable and animal food supply from unnecessary exposure to materials that can adversely affect short as well as long-term human health Studies in the Jane Adams' Hull House in Chicago ca 1900, explored these issues in the surrounding poor immigrant working community [20].

The early prime driving force for design optimization was maximizing production while minimizing costs. Early on there was little to no ESOH regulatory consideration. Also, a limitless supply of feedstock left bartering for the cheapest feed source the primary concern of cost controls. As urban populations grew, the industrialized world grew more dependent upon finished products and the importance of securing raw materials and energy sources ever more important. (Countries still consider it in their "national interest" to maintain free access to both, and the great powers attempted to address this at the end of World War II through the UN, NATO, SEATO, and other enlightened peace-oriented methods to replace colonialism.) Around 1900, the need for standards to protect human health was identified and Food and Drug Administration (FDA) came into existence to protect against "snake oil" and other patent medicine fraudsters. This arose in part from a population feedback as people became sick or died and these events were reported or they simply chose other manufacturers products, based on efficacy. Product labelling requirement informs industry which in turn responds via competition. The USDA was of particular importance for food and FDA for drugs, but all were aimed at consumer protection, though not yet with sustainability in mind. Both were signed into law (Pure Food and Drug Act of 1906) by the Republican President Theodore Roosevelt, as a result of crusading journalists, e.g., Upton Sinclair [19] and others.

1.8 Improving Classic Process Engineering Design: The Key to Success

A good process diagram, material and energy balance, coupled with engineering economic cost analysis is the stock and trade of the ChE process design engineer. Originally, drawing a picture around the plant, aka battery limits, was sufficient. Then it became obvious, as in refinery operations, that a large amount of energy in the form of heat was lost to the environment or worse yet had to be balanced by means of wasteful, artificial cooling, wasting large quantities of water from rivers, lakes, streams, and ground, simply to maintain safe and optimal operating conditions. Designers soon engineered inside battery limits (ISBL) heat exchanger networks (economizers) to recover hot product heat by exchange with cooler feedstock. On the material balance side, early process wastes were not recovered due to a lack of economic viability or suitable methods of separation and purification or, for that matter, waste-handling regulations. More recently, an unexpected undesirable consequence of P2 and hazmat regulations is over-restrictive definitions on a manufacturer's reuse of waste as a feed source, creating an impediment to recycling.

With the identification of adverse health effects resulting from chemical and related manufacturing, laws were enacted to reduce land, air, and water emissions as well as worker exposure. Further, it became apparent over the last 50 years that natural resources, without husbandry, were unsustainably depleting. This includes but not limited to: trees, topsoil suitable for farming, pollinating species, e.g., bees. Large-scale manufacturing operations typically operate in continuous, 24 h 365 day/year mode and must have access to a constant flow of raw materials and energy sufficient to process them. More often than not, these resources are nowhere near the battery limits, so a bigger circle had to be drawn around the process to account for these variables (1000s of miles, perhaps).

The EPA and OSHA came into being in 1970 in response to public health concerns and at once helped define/refine the concept of material and energy emissions for manufacturing that was both health and cost based. One of the principal environmental laws finally prohibited the placement of toxic waste into landfills, but has still failed to adequately address the concept of recycling these buried "waste mines," but this surely needs to be worked out. At least many municipal solid waste (MSW, or garbage) landfills have perfected methane recovery methods; but, in the future, these materials, organic as well as inorganic, should be kept out of landfills completely, or at least wherever practicable, and reused.

Disparate laws within the EPA that set recycling against waste disposal often result in contrary rules that discourage the safe re-purposing of such former waste products, as we shall see are anathema to SE. This can be addressed by preventing waste creation in the first place. Also, EPA regulations state that a material is not a waste until the generator declares it so. This fact should always guide proper recycling efforts to prevent "sham recycling," or illegal hazardous waste storage beyond 90 days.

✈ There is money to be made in P2—US Air Force example at Keesler Air Force Base, Biloxi MS of separating glass into various colors, newspaper, building cinder blocks into simulated coral reefs to support gulf ocean fish breeding, metal into steel and aluminum, plastics by type, etc. Treat MSW as a commodity and you can make money! ✈[1]

1.9 Interconnectedness of Everything

We now have a better and constantly evolving understanding of how all life on the planet is connected, and how the various species of flora and fauna interact and must stay in balance. This balancing act has short-term cycles easily observed when overfishing or hunting takes place, and longer term cycles we have yet to discover and/or fully appreciate. The long-term cycles, e.g., mini-ice ages and global temperature change, have been well documented over tens of thousands of years in the ice and fossil records. Modern interpretations of these cycles have become politicized, as any new rules might cost industry money to implement. Industry is accustomed to event horizons of 2–3 years at most, and the financial impacts of long-term effects, e.g., glacier disappearance become more difficult to predict accurately. Greenhouse gas legislation is a good example of such polarizing positions that will require international resolution.

Earth cycles: Our very existence seems dependent upon this balance on earth, which itself receives an *assumed* permanent supply of solar energy from which energy is drawn. Vegetative life forms convert this energy into useable matter upon which our very lives depend. They do this not by magically creating matter, but rather by reformatting existing atoms of C, H, O, N, S, into plants which in turn are eaten by herbivores. Animals breathe in O_2 and exhale CO_2 and H_2O. Plants and other life forms take in CO_2 which they use for growth and exhale O_2. A perfect balance, only if we observe the long-term connectivity which again is an overarching purpose of adding SE into the design equation—maintaining steady state for planet earth.

As part of its mission, EPA has purview over the protection of species. The connectedness of all things will require a global approach to this. SE design is apolitical. If practiced, it will reduce material and energy use as well as total cost, with additional unexpected positive consequence of reduced CO_2 footprint.

[1]✈ Author's comments within these symbols.

1.9.1 Quality Management Approach for Complex Interconnected System Operations

An argument can be made that the design engineer must draw his/her circle around the planet to assure the long-term success of any planned as well as existing manufacturing facility. Perhaps the planet is a bit ambitious, but certainly guaranteeing constant flow of energy and feedstock into the manufacturing process while minimizing adverse external impacts gets us closer to a definition of the goal of sustainability engineering. Quality practices are key to managing complex systems. QM also demands fact-based data and so is inherently apolitical. Someone said, "You are entitled to your own opinions, but not to your own facts." (Senator Daniel Patrick Moynahan)

1.9.2 Considerations of Sustainability Engineering

Scientists are primarily concerned with understanding the physical world around us, the "why" of things. Engineers main interest is in "how" to make things happen (the why comes later), so for example, while it may be of little short-term interest to the engineering designer as to why the earth's temperature is increasing, it is, however, of paramount importance in the design of heat exchange equipment and water cooling towers to factor in additional, more costly, cooling capability in climates whose temperature is increasing regardless of the cause. In a similar manner, rising sea levels impact coastal architecture and builders' designs, lowering groundwater table impacts farming and drinking water supplies, all requiring sustainability-based attention. All the while, designers must factor in manufacturing plant project lives of 10–20 years that must operate 24/7/365, making these decisions all the more critical for both sustainable and economic success. Indeed, if done properly, the addition of sustainability engineering to classic process design engineering should lead to enhanced long-term project profitability.

Investors are naturally and rightly risk averse. But as our understanding of SE design technical issues comes into sharper view, the enhanced SE engineering and economic evaluation tools will allow for better bottom-line decision making and improved long-term profitability. But this is the same thing that happened with industry adjustment to the ESOH laws that were introduced largely with the formation of EPA and OSHA in the 1970s, and their various adjustments, additions, and modifications.

Unbridled population growth and uninformed capitalism in a desirable living environment like California, for instance, is presently leading to severe water shortage for vital human and agricultural use as well as the one of the worst wild fire seasons in recent history, really only because earlier decisions were not tempered by SE design–based input. Overarching SE guidelines, perhaps even legislation, could go a long way toward assuring that such decisions do not hinder economic growth, but rather, factor in considerations necessary to sustain that growth. The costs are small on a per capita basis of end users and will protect investments long into the future. As an example, building codes are relatively new in the USA. The Underwriters Laboratory (UL) came into being in part to provide life safety standards for the nascent electric lighting industry in the 1880s following the Great Chicago Fire of 1871. Today, no one would purchase a building that was not to code in all aspects, and this has not hampered the honest business developers' need for relatively short-term profits; and insurance companies will be pleased to reduce safety-related accident claims payments. SE can work, if the present system is uniformly and fairly modified. No need to throw out the baby with the bathwater. Greed still works, if all consequences are known to all affected parties. Healthy business competition is still a good thing if all play by the present and developing rules going forward.

Present day economic considerations, typically centered on the 2–3 year simple breakeven window, could be tempered by use of 5-year plans that allow companies to reassess directions; with supporting IRS Tax Code Modification.

For sustainability engineering to be truly practiced, some new international laws and/or agreements may be needed. In the USA, NAFTA, e.g., already outlaws the transfer of US pollution that may arise from shifting manufacturing away from the USA. (Unfortunately, this has not been really enforced, putting compliant US manufacturing at a competitive disadvantage, but a good example of what might be needed to avoid medium-term non-sustainable operation.) Europe first adopted then ran ahead of the USA on ESOH regulatory development. China is now coming to grips with the limitations of its centrally planned economy and largely unbridled ESOH regulatory environment which has led to severe pollution as well as massive explosions and contamination of food and other products, e.g., formaldehyde in wallboard, reminiscent of pre-FDA USA, ca 1900! These imported Chinese wallboard materials were sold into USA and other markets and will cost the Chinese billions of dollars in litigation now to remedy. The recently signed Trans-Pacific Partnership is supposed to address these issues.

1.9.3 Combined Manufacturing and Power Generation: "The Only Thing New in the World Is the History You Don't Know!" (President Harry S. Truman [6])

EPA describes combined heat and power (CHP) typically in regard to energy production and heating. Power generation has a very low thermodynamic efficiency. So, Harry was correct, there is nothing new here.

But there is a better way that is discussed in Chaps. 4 and 7 on SE efficiency and SE tools that focus on combined integrated power and chemical production. This is the key to optimizing energy and material utilization, not just across one process, but by combining disparate processes together to squeeze every last drop of material and energy resources, while minimizing waste production and ESOH effect. We will examine this in greater detail in Chap. 8, SE examples.

1.10 New SE Approach: Integrated Power and Processing Plants

The chemical process industry (CPI) has fewer energy utilization limitations than the power generation industry, as both low and high quality heat can be used in the CPI if enough diversity of operations exists. Refineries have long since made use of these principles ISBL, but the CPI can do the same by combining unrelated processes together to minimize waste generation and maximize energy utilization. Integrated power and processing plants in the CPI can learn much from refinery design and operation.

1.11 The Gasifier as a Swing Unit Operation (SUO)

Gasification [8] can be employed to produce feedstock for the production of fertilizers, steel, and fuels, to name a few. Recent fracking-induced natural gas supply increase provides logical starting point for many diverse products, but this will require a new paradigm, as companies will need to either develop or team with existing chemical manufacturer's possessing sufficient proprietary knowledge. A wealth of material feedstock also exists in MSW landfills. Methane from some of these facilities is already being captured for energy but metal values could also help reduce the MSW footprint while diverting oil from chemicals production or vice versa and reducing new mining requirements. This and other technical toolbox examples are discussed in Chaps. 4 and 7. As will be seen, President Truman was correct in that there is not much new, but we will be improving these old standby technologies, at least for efficiency, or replace as necessary, to better serve the SE-designed operation. Gasification is a sort of catch all for keeping all organic material out of the landfill.

1.12 Regulatory Updates for SE

Regulators will need to update rules to allow processing of hazardous waste within SUO's that can convert them into finished products. In my readings of USEPA regulations, I found that a material is not a waste until you say it must be applied here. This means that it is perfectly acceptable to place the end of one process's pipe, one that used to be considered waste, into the feed end of another process. The CPI already knows how to take toxic materials, e.g., maleic anhydride, and produce at very high purity nontoxic polymers to safely augment crop growth and will not need to be hindered by outdated TSDF rules so long as they are not handling or disposing of wastes. The transfer of such materials from one company to another, however, may require regulatory updates to avoid triggering sham recycling laws. We examine this in Chap. 10.

1.13 The Whole Point of It All

This book approaches the topic of sustainability engineering from a holistic approach. Although the primary focus is on the chemical process industry, we examine some very important applications in the other engineering disciplines, namely bio, civil, electrical, environmental, industrial, agricultural, and medical. To me, the subject of SE revolves around the safe and economical implementation of process designs that are used to turn large quantities of raw materials into useful finished products for human or animal use and consumption, safely, economically, and always sustainably.

Sustainability can mean different things, but here it refers to the perpetual availability of the given material and energy resources based on either recycling/reuse/repurposing of other materials. It can also refer to inexhaustible supplies, e.g., solar energy, human waste water, human garbage, and human or animal wastes. By definition, sustainability engineering will include the husbandry of these resources in a manner that produces minimal negative impact elsewhere as a result of unintended consequences of resource collection and use for manufacturing. (No pollution exporting!) A trade policy to protect US manufacturers who do comply would help level the playing field a bit.

1.13.1 Great Challenges and Opportunities in Sustainability Engineering

- Defining sustainability: fact-based decision-making to bridge well into the future—easy
- Capturing useful and reliable data—not so easy, but eminently doable
- Tie the whole thing to ESOH and performance-based contracting

1.13.2 Sustainability Engineering Approach

- A bridge to tomorrow
- A "how to" primer/review for the designer, plant engineer, and academicians
- The sustainable engineering thought process and approach

1.14 Summary

The subject of pollution prevention and "sustainability" has been a long interest [5, 9–11] of the author.

As noted, there exists a vibrant literature on the subject of process and product design [12–18].

There will be no attempt at reinventing that wheel, but rather to build upon it. "Sustainability engineering" as presented here is merely an element of continuous improvement to the existing body of work covering the subject of design and operations in the CPI. It is much needed if we are to move safely and economically into a world of ever-increasing demand in the face of diminishing resources.

References

1. Carson R (1962) Silent spring. Houghton Mifflin, Boston
2. Perl JP (1990) Coming clean—superfund problems can be solved. US Congress Office of Technology Assessment, Washington, DC
3. USEPA. www.epa.gov
4. Total Quality Management—American Society for Quality (ASQ). http://asq.org/learn-about-quality/total-quality-management/overview/tqm-history.html
5. Perl JP (1994) A quality approach to pollution prevention and waste minimization through quality management, environmental models and training, Presenter and Co-Chair. AIChE Chicago Section Fall Symposium, Chicago, IL
6. Miller M (1974) Plain speaking: an oral biography of Harry S Truman. Berkley, New York, p 26
7. Doctor John Snow blames water pollution for cholera epidemic. http://www.ph.ucla.edu/epi/snow/fatherofepidemiology.html
8. Higman C, van der Burgt M (2008) Gasification, 2nd edn. Elsevier, Burlington
9. Perl JP (2007) "Integrated energy and material conservation in modern chemical plant processing", presented at 11th Green Chemistry and Engineering Conference, American Chemical Society, Capital Hilton Hotel, Washington DC, June 29, 2007
10. Perl JP, Peters RW (2007) Savings and optimization: chemical process industry. In: Capehart B (ed) Encyclopedia of energy engineering and technology. Taylor and Francis, Boca Raton
11. Peters RW, Perl JP, Peters RW (2007) Savings and optimization: case studies. In: Capehart B (ed) Encyclopedia of energy engineering and technology. Taylor and Francis, Boca Raton
12. Turton R, Baillie RC, Whiting WB, Shaewitz JA (2009) Analysis, synthesis and design of chemical processes, 3rd edn. Prentice Hall, Upper Saddle River
13. Seider WD, Seader JD, Lewin DR, Widago S (2009) Product and process design principles, 3rd edn. Wiley, New York
14. de Klerk A (2011) Fischer-Tropsch refining. Wiley-VCH, Weinheim
15. Towler G, Sinnott R (2008) Chemical engineering design. Elsevier, New York
16. Peters MS, Timmerhaus KD, West RE (2003) Plant design and economics for chemical engineers, 5th edn. McGraw Hill, Columbus
17. Douglas JM (1988) Conceptual design of chemical processes. McGraw Hill, Columbus
18. Cussler EL, Moggridge GD (2011) Chemical product design, 2nd edn. Cambridge University Press, Cambridge
19. Sinclair U (1905) The Jungle
20. Adams J, Occupational Therapy at the Hull House, Chicago

Chapter 2
ChE Sustainability Engineering Design Approach: Bread and Butter

Until relatively recent times, the overall approach to process engineering design considered what is needed, what can be spent, how much can be made, and how fast can an investment be recovered and profits begin to flow. In a world of overabundance, unlimited space, and raw materials, this was sufficient. Modern SE design however, must also consider safer alternatives to the desired product, permanent availability of raw materials, with strict, primary emphasis on renewable resources, and Environment, Safety and Occupational Health (ESOH) compliance and protection for the public as well as industrial workforce.

The basic cost estimating, technical requirements determination, and methodology of design have not changed, e.g., the manner in which equipment is selected and sized. But CAD-aided design methods, e.g., Aspen plus, Hysys [1] and others, e.g., UNISYSM can greatly simplify, through automation, the level of effort and often provide a greater degree of accuracy in design. In some cases, dynamic, real-time plant operational data can be used to design, build, and continuously improve the computer-based manufacturing model. Interactive dynamic modeling can also provide a sort of self-tuning operations mode that corrects the initial production model algorithm for improved process control. This, in effect, can be viewed as a sort of facility "calibration" to augment initially assumed physical properties–based modeling, and can be used in troubleshooting to identify process changes such as fouling, catalyst efficacy reduction [2], even material and energy leaks and losses. All of this might be considered part of what is now referred to as artificial intelligence, or AI. If so, the CPI was way ahead of that curve.

2.1 Classic Process Design Steps

Problem Statement—The Design Basis
Flow Sheet Basics—The Block Flow Diagram (BFD)

J. P. Perl, *Sustainability Engineering for Enhanced Process Design and Manufacturing Profitability*, https://doi.org/10.1007/978-3-031-52363-2_2

Material Balance
Energy Balance
Refined Flow Sheets—The Process Flow Diagram (PFD)
ESOH—Preliminary
Conceptual Process Control
Equipment Sizing and Selection
Economics—Rough Stage Gate Go–No Go
Detailed Design Engineering P&IDs
Final Construction Design Package
Design/Build/Commission/Operate

For all newly developed chemicals, prior to going much beyond the material balance and even before, a preliminary ESOH review (Chap. 6) should be undertaken to avoid later problems. A preliminary USEPA premanufacturing notification (PMN) with EPA's Ecological Structure Activity Relationships (ECOSAR) review (Chap. 5) should also be done just after the initial design basis has been proposed to evaluate permitting and operability, as well as to identify less toxic alternatives. This really should be a part of all *New Product Design* activities prior to process design itself. This is an operating plant EPA requirement anyway and might also lead to a better, safer, less expensive product.

After the material and energy balances are completed, comes flowsheet based equipment selection and sizing. This step is highly dependent upon accurate physical property modeling. An increasing dependence on modeling programs means great scrutiny and understanding regarding the selection and use of physical parameter selection within those programs.

From an equipment list, a purchased cost is determined allowing a preliminary factored estimate of total installed costs (TIC). Then, rough economics will help in a stage gate go–no-go review. During the equipment selection process, process efficiency must be taken into account or risk long-term economic loss (Chap. 4). Selection of least-cost equipment without consideration of long-term operating costs might look good upfront but can be economically disastrous over the actual project life.

Many different variances on this classic design mode and other similar ones are already in a high state of development within the engineering design community. To this, we will add elements to support an enhanced sustainability engineering (SE) mode, which in many respects is really just a finer look at the overall process and recognition of a need to tighten upfront estimates while looking out well beyond the traditional calculation borders.

2.2 Sustainability Engineering Unified and Integrated Process Design Elements Module

- Fitting Power and Processes Together Optimally
- Cross Platform Material and Energy Balance

  - Using Renewable and Recycled Resources
  - Sustainable Resource Optimization
  - Coupled Heat and Mass Balance
  - Minimize Transportation by Plant Collocation

- SE-Rated Equipment Efficiency Standards
- Optimize Across Divergent Profit Centers

  - When coupling two or more processes, this must be considered

- Coupled Heat and Energy Production and Utilization in the CPI

  - Small- and Large-Scale Generation

- Environment, Safety, and Occupational Health Considerations—Check this first!
- Resource Optimization and Recovery Review

2.3 New Core SE Design Paradigm

As noted, SE design must be seen as a fundamental improvement to an already well-developed and successful process design methodology. Just as ESOH regulations of the 1970s led to safer, more economical, less damaging processes, and ultimately more profitable operation, SE design will have the same positive effect once properly incorporated. This is why the 3M company motto from the 1990s of "Pollution Prevention Pays" still prevails, as it allowed the company to put a finer magnifying glass on the chemical manufacturing process and in the balance, found improvement all around.

This finally put to rest the notion of doing things the same way "because that is the way we always did it." It also demonstrates why quality management (QM) (Chap. 9) is so important. My Air Force experience in particular also made clear why training is so important, not just once for an individual, but in recognition that there is a constant flow of new individuals. This fresh crop cannot be injected with cumulative corporate knowledge of those leaving and so must be trained through established procedures and practices. The same phenomenon occurs in industry and this first showed up magnified in regards to process safety management (PSM). PSM is the OSHA requirement that was borne out of one too many industrial explosions and events causing death and injury from preventable accidents.

So, training is a primary checklist item for quality management and includes both generalized quality management training and the specific technical training components required for the job at hand. In keeping with QM principles, the technical training elements will need to be updated as we turn seemingly random unknown unknowns into knowns and to accommodate any and all systematic changes. Process safety management (PSM) recognizes this as "management of change" and is quite proscriptive in its treatment. This is why I added a chapter (Chap. 9) on Quality Management and added a way to take input from all involved with production, creation, etc., to assure optimal outcomes.

2.4 Process Technology Efficiency: Key to SE Success

An important aspect of successful SE design lies in the selection of technology with as high as practicable efficiency. Because of interconnected power and process and the stronger degree of coupling that arises, there is much need for improvement here. Simply making something a bit more efficient has always been a design objective and will no longer be enough. The process equipment optimization needed for 5-year projects is far less demanding than that required by SE. In addition, the focus on early economic payback while important can have the unexpected consequence of de-selecting more efficient technology in order to minimize initial capital expense cost (CapEx) outlays. This in turn leads to long-term economic losses in the form of increased energy use. Once low-efficiency equipment is in place, removing and replacing it to save energy can be cost prohibitive. Remember the 5x rule, i.e., every $100 of equipment will have a total installed cost (TIC) of around $500, so do it right … right now!

Selection of high-efficiency equipment can also boost a classic process design into the SE realm. As an example, highly inefficient incandescent lighting went to better fluorescents and now to the light-emitting diode (LED). As LED power electronics improve, application of LED lighting will greatly reduce power consumption to the point where solar photovoltaic level (PV) might be a sufficient energy source for many applications. Even when using nonrenewable resources, more efficient chemical process equipment might also serve as a bridge to SE's future.

2.4.1 Energy Conservation and Efficiency Improvements: An SE Extender

For some time now, electric power utilities have encouraged conservation. This form of "enlightened self-interest" works to the consumer's advantage by saving on annual costs. It is also a brilliant method for utilities to avoid the cost of building new generation capacity, a very expensive proposition per kW added.

Production facilities commonly already do what they can to reduce energy consumption for obvious reasons. However, for SE, it is imperative that both continuous improvement in energy consumption and selection of highest efficiency power equipment for new construction take place. Remember, short-term selection of least capital cost equipment will lead to a lifetime of excess, unrecoverable power costs.

2.4.2 Material Conservation and Efficiency Improvements

Owing to the regulatory nature as well as contract delivery requirements of chemical production, changes here are a bit more involved than those for energy. Still, it is an important checklist item for continuous optimization of long-term production facilities. New product design is discussed further in Chap. 5.

2.5 A Note on Process and Product Design Modeling

Although beyond the scope of this book, some SE-related comments are appropriate. Efficiency and thermodynamic models are often coupled together. For large-scale complex plant production and operations, modern process design incorporates computer-aided design (CAD) programs, e.g., Aspen Plus, ChemCAD, and ChemStations, HYSIS, VISIO, and AutoCAD to name a few of commonly available tools. But the process modeling tools all rely on thermodynamics packages to estimate physical properties, unit operations and reaction engineering.

Largely through the work of Prausnitz et al. at Berkeley e.g. [3] and others, the advances in our understanding of molecular thermodynamics have moved us well beyond the simple ideal gas law and other earlier physical–chemical ones, e.g., viral coefficients, van der Waals, and reduced properties bring computer modeling to new heights of localized plant design. These design tools can also be used to live-model optimize existing plants as well.

This is a prime example of how best the academia–industry feedback loop can work. But any computer-modeled process design that is based on incorrect thermodynamics will be suboptimal at best, incorrect at worst.

Similarly, incomplete engineering economic evaluations without SE considerations will be suboptimal in the same way that fixing ESOH after the plant is built is far costlier and potentially fatal economically. Actually, this is all about reducing unknowns to knowns and then applying accepted design principles to address them.

This may seem squishy at first, but it is all about optimizing the existing and planned plants with respect to a larger boundary than we're used to. The money folks will need to understand how to factor in loss of a feedstock that comes from overharvesting, overmining, etc. Another paradigm shift driving this is an international movement away from a military-based national interest policy, i.e., maintaining fair market access to external oil supplies and critical metals. These considerations will probably never go away completely, but they can be reduced by application of SE approach that promotes self-sufficiency. And do not forget to factor in the cost in blood and treasure of war into the overall economic equation.

Sustainability analysis presents a promising new challenge. The uncertainty of estimating material and energy availability will give a "best" range SE solution that might lie between say 15 and 25 years, i.e., very flat. Business, in general, is not very good at looking beyond 2–3 years for breakeven investment purposes. There are many variables at play such as noneconomic-based war, economic downturns, acquisitions and divestitures, and failure to comply with ESOH laws between countries, to name a few. Still, the loss of the resources necessary for production is so financially devastating to large-scale operations that methods to factor in sustainability will move to the forefront. The government can have a role here and already does when it comes to things like the strategic petroleum reserve, helium storage, and metals critical for national defense.

Our world would be much less safe without ESOH and FDA and other protective regulations which have provided great benefits to people without hampering production or when viewed through a large lens, profitability. There may be a role for

SE regulations of some kind. Taken as a whole and in the long view, all of these can work in concert to improve our overall quality of life which includes not only health but also a full economy to ensure its continuation … true continuous improvement! P2 paid and so will SE!.

A few things should always be present in the SE checklist:

1. Energy evaluation (resource availability)
2. Materials evaluation (resource availability)
3. Renewability life span (even the sun will burn out someday!)
4. Integrated power and chemical production potential
5. Law of unintended consequences (LOUC) review
   (a) Will it hurt something elsewhere—assign cost to this

2.6 New Overall SE Design Approach

The following sections examine various SE elements and how they begin to fit together to enhance classic process design. Later chapters drill down into programmatic details followed by some SE examples.

2.7 Integrated Power and Process Design Engineering Elements: Fitting Together Optimally

In an attempt to provide an overview and methodology for SE, the following is presented. For all of my contemporary fellow engineers, the core process design methodology has not changed. Rather, a magnifying glass on material and energy utilization with an eye on efficiency improvement and resource conservation emerges that is wholly within the domain and province of the professional design engineer or those responsible for plant operations and improvements. Key also is the integration of power generation as well as the addition of disparate processes necessary to provide a material and energy balance closure while minimizing waste and maximizing resource utilization. An SE rating system is proposed to assist in delineating degrees of SE in the following checklist.

2.8 A 40,000 Foot View: An SE Design and SE Rating Approach

1. Classic design basis review—Already works well, just needs continuous improvement (who does not?)

2. SE process integration review—Fundamental improvement
   (a) Material exchange
   (b) Energy exchange
   (c) Facility power integration potential
   (d) Use of SE high-efficiency technology
   (e) Early involvement of all design engineering disciplines and team members
   (f) Operability and interoperability review
      - Maintenance, Management, and Operations
   (g) ESOH and alternative chemistries review
   (h) Reformat or reuse elsewhere irreducible process waste
3. SE material and energy sourcing and rating—Fundamental improvement
   (a) Good
      - 25-year renewability review
   (b) Better
      - 25- to 50-year
   (c) Best
      - 50 to 100+ years
   (d) 100+ years = Golden (or is it platinum now?)
4. When the above material balance (MatBal) and integration timelines extend past the economically required facility production life then this yields → successful SE Design
5. Consider developing an SE design sustainability star rating
   (a) Most States already have a Governor's Pollution Prevention Award

The classic/SE design modules loop around until solution convergence is met. This is really no different than the previous SE. Various stage gate accuracies are used in the same traditional manner, i.e., conceptual design for go–no-go decisions, ±25–50%, followed by ±25%, then ±10% all the way to Issue for Construction (IFC) drawings. Stage gates are often employed at various points in the design process as fact-based go–no-go decisions. The common ones are ±50%, ±25%, and ± 10%, leading up to final firm fixed bid estimates. Each stage gate adds an additional layer of detail review and often uncovers design flaws or modifications needed prior to final design and construction. Final economic goal is to minimize change orders and cost overruns. An overall SE design process referring to Sect. 2.1 is depicted below (Fig. 2.1).

The civil engineers (CE) commonly design infrastructure in the 50–100 year range; so, perhaps this is too short a time frame? Does it give the CE time to develop newer, better, and more economical replacements? I would say it does, given the advances seen in building materials and methods over the last century.

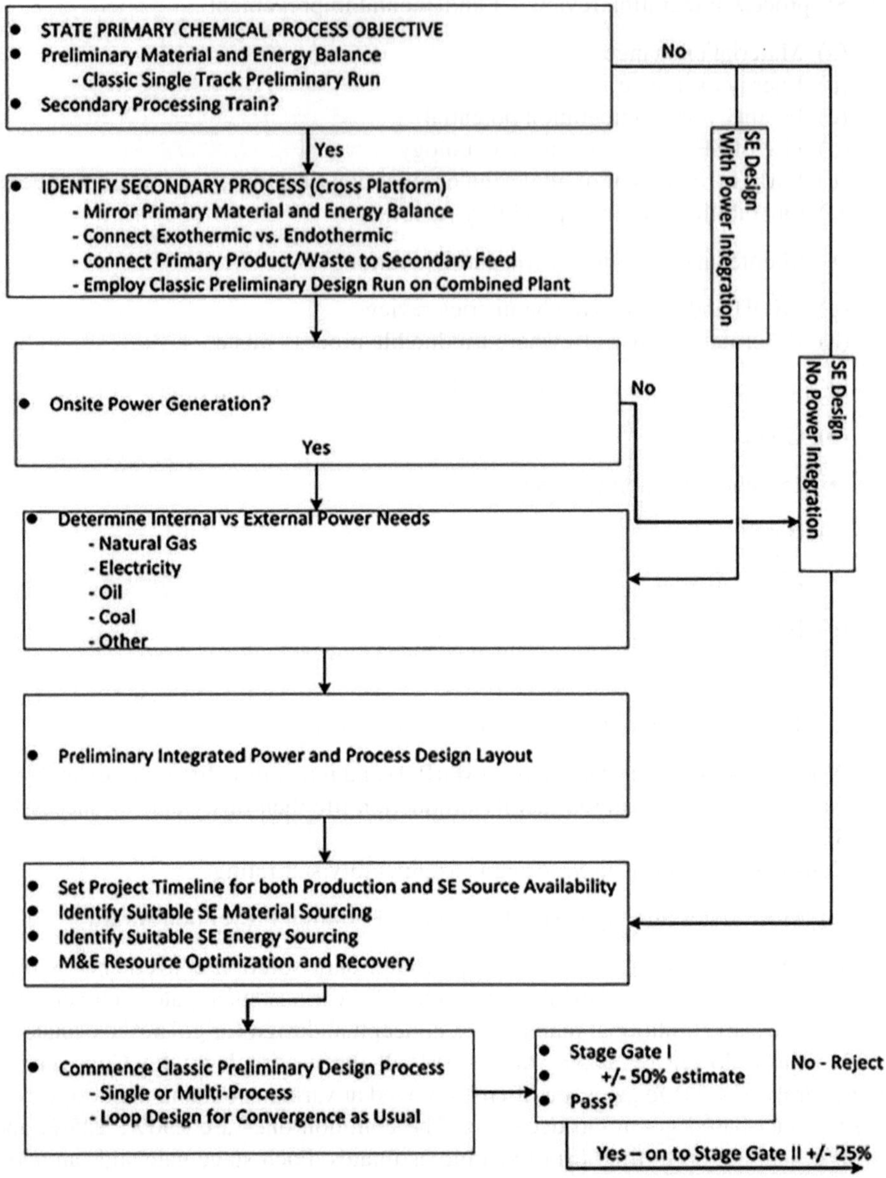

Fig. 2.1 Sustainability design methodology flow chart: All steps are preliminary until final classic design

2.9 Prior to SE Design

Pre-1970—First, a design basis is established regarding the desired project outcome. This is done in a traditional manner with a simple block flow diagram, material and energy balance around the blocks, and preliminary cost estimate. In 1970, EPA and OSHA were established by federal law. Around 1979, a rule requiring new chemicals to be go through a premanufacturing notice (PMN) EPA submittal was added. Also EPA added land disposal bans severely limiting burial of toxic materials and leading to a more pollution prevention approach.

By the late 1980s to early 1990s, this led to the first modern fundamental design paradigm shift, namely the incorporation of environmental and health and safety laws as an upfront consideration in chemical process design to support both new and planned expanded manufacturing operations. Federal laws were necessary to the establishment of fair, across the board manufacturing rules. Later, pollution prevention goals and guidelines were established but by that time, industry had already started working proactively to do this, as it helped assure the minimization of future costly legal responsibility arising from controlled as well as uncontrolled releases and excessive worker exposure. It is worth noting that both civil and criminal (willful) penalties undoubtedly played a role in increasing compliance. It is also worth noting that all of this has actually saved money in the long run through improved process reliability and human health.

2.10 New Sustainability Approach: Consumer Driven: Process Required

As recycling became a more public thing, Industry got the message that consumers were interested enough to "vote with their feet" and shop for environmentally friendly products. This includes everything from food to electronics and in particular household cleaning, homecare products, etc. The topic of global warming became a presidential campaign issue in 2000 and 1988 with the publication of Al Gores' book, *Earth in the Balance*. It is still a major political issue. Since its inception in April 1970, Earth Day has been institutionalized and children in kindergarten are encouraged to recycle to save the earth. I have found teaching senior chemical engineering process design with an eye toward SE was quite easy and taken as a positive challenge by the students.

Sustainability describes human activities necessary to life that can be accomplished with minimal to no adverse effect on the external environment, locally as well as globally. Interest in identifying and developing renewable resources are key to the success of sustainable anything. Add to this the ever-increasing industrial competition to sell goods and services and this creates a perfect storm of positive, unintended consequences, convergent on sustainability.

The integration of onsite power generation into production will propel SE to higher levels. Power, coupled with external disparate process integration opportunities, is the primary objective. The key here is matching exothermic with endothermic processes, as well as product/waste from one process with feed to another, sometimes unrelated disparate one. Again, in the preliminary design phase all that is necessary is to get a rough idea of what might work. Remember, disparate means just that, i.e., processes that normally would not be put together. These options are racked and stacked against internal material and energy recycling options. A refinery [4], for instance, has many options of energy recovery and so might already be in a high state of localized SE design. This is an internal fact, however, as the facility is still dependent upon nonrenewable petroleum resources.

With natural gas from fracking one can produce natural gas liquids (NGL) and the refinery of the future might eliminate gasoline production entirely and focus primarily on petrochemical production, which in turn might place it into the best category of sustainability simply by extending oil resources out well beyond 100 years. This might even include reformatting manufactured by-products as feed into the refinery (may require EPA rules change). Of course, the fracking gas is nonrenewable itself, but may serve to bridge to the future, while reducing CO_2 footprint when done through SE.

The second new SE step proposed here involves a rating of the material and energy sources in terms of the above criterion. In the past, the only assumption made was regarding raw material specification. For SE design, however, identification of raw materials, purity, quantity, and assured availability is reviewed simultaneously. The list in Sect. 2.8 arbitrarily delineates four levels of SE. Note that the goal of SE design is to provide fundamental improvements to an existing design. Moving the timeline out at least last 50 years will, hopefully, provide a bridge to new SE technologies. Solar photovoltaics, algal oil, to name a few, are making great advances that will someday remove crude oil entirely as a source of motive fuels and move it more squarely into sustainable, recyclable chemical production, e.g., plastics and other petrochemicals.

These SE criteria taken together force a tighter material, energy, and technical evaluation and focus layered on top of the traditional design process. The same effect took place when ESOH regulations came into being that ultimately led to safer, better, and more profitable manufacturing methods. With SE, however, instead of assuming that raw materials and energy will be available forever or worse yet, not assuming anything and trusting only to market forces, this new paradigm will lead to increased productivity, and markets as consumers continue to demand sustainable products and producers realize significant long-term cost savings.

SE will also help the Industry get back to *industrial R&D* to develop more efficient, cost-effective, sustainable methods in a wide variety of categories as suppliers compete to provide SE technologies. Furthermore, an emerging academic research focus on "new product design" will compliment and hopefully reestablish and reinvigorate the industry–academia interface. Largely through government funding, academia has become much more theoretically oriented in recent years, and the

Industry can and should tap into that at least to jump start new improvements in kinetic and molecular thermodynamic and interfacial properties modeling just to name a few. Done properly, this will be profitable through increased mechanical efficiency and hence productivity.

As we tighten up SE design criterion, integrated designs in particular will need higher efficiency technologies. As the degree of interaction becomes more complicated, it becomes necessary to have less variability in efficiency, e.g., in turn down ratios for integrated equipment across disparate processes. This will shed new light on and encourage development of *SE-branded technologies* to support the new SE design paradigm. Chapter 4 on the "Efficiency of Everything" highlights that key SE element.

No one individual can integrate ESOH with process as well as power generation within the SE umbrella. Coordination of waste into feed and external energy transport might be new roles for the plant environmental coordinator. Again, this is why I place so much emphasis on quality management programs including QM training. This is not commonly taught in engineering schools so industry may need to step up as needed just as it has in response to process safety management.

Existing equipment and technical standards development may also need tightening. Some sort of an SE-Star Award for meeting the highest level of renewable resource utilization as previously discussed would help designers identify SE technologies. Individual States already commonly give Governor's Pollution Prevention Awards, so the awards and recognition delivery mechanism already has an appropriate home all around the USA.

2.11 SE Design Team Ground Rules: Quality Management Based

With the addition of integrated power and disparate chemical processing on the same site, there are fewer degrees of freedom available to a successful SE-designed facility. Not only are the technical components in need of fine-tuning, but it also will be more important than ever before for the complete design team to include suppliers, engineers, and multiple end-user clients to assure success. This is not to say everyone is involved at all stages or at the same depth, but there will be no other way to provide process guarantees that industry ultimately requires for all financially successful engineering, procurement, and construction (EPC) projects.

As in classic design, stage gate decisions are made at key points to allow for more refined economic estimates that will, of necessity, sharpen SE review where material and energy resources are involved. The steps then loop back around until a final design and issued for construction (IFC) drawing design packages are produced, just as they are in contemporary engineering design projects.

2.12 Slightly More Detailed Sustainable Engineering Process Design Approach

Sustainability engineering design approach by the numbers:

1. Problem statement
   (a) The primary design basis—what are you trying to do
   (b) Quick sanity check on ECOSAR and alternate chemicals
   (c) Process and power integration basis
   - Additional SE power generation?
   - Additional secondary SE disparate process?
2. Flow sheet basics
   (a) "Draw a picture"—back of the envelope
   (b) The block flow diagram—connecting things together
3. Material balance (MatBal)
   (a) This is first and starts the calculations, design-basis driven
   (b) Start of materials SE—must assure future resource availability
   (c) Process waste reutilized or converted to energy
4. Energy balance
   (a) Always coupled with material flows, so this comes after the MatBal
   (b) Start of energy SE—must assure future resource availability
   (c) Coupled to onsite power production, where applicable/possible
5. Refined flow sheets
   (a) The process flow diagram
6. Conceptual process control
   (a) Identify basic needs and challenges
7. Economic, ESOH and sustainable resource optimization
   (a) SE-driven check of impacts on human, animal, and plants
   (b) SE-driven review of all material and energy sustainability
   (c) Classic engineering economics
   - Grant and Ireson
   - Lean accounting (Myerson, Anthony industry week)

   (a) PMN and permitting
8. P&IDs—Architects and Engineers (A&E) endpoint
   (a) Detailed drawings that show all piping, material, and energy and controls

   (b) Bid specifications

9. Design/Build/Commission/Operate—Engineering, procurement, and construction (EPC)

   (a) Detail design engineering
   (b) Issue for construction drawings
   (c) Construction
   (d) Startup

   - Nameplate commissioning
   - Steady state operation
   - Client handoff

These taken together are considered to comprise sustainability engineering-based design and require simultaneous consideration and optimization.

2.13 Tough SE Nuts to Crack Include

1. Renewable raw materials—New to SE

   (a) External impact of diverting any raw material into a given process
   (b) Legal aspects

2. Renewable energy source—The sun is it for now in the strict sense
3. Overall energy integration—New to SE

   (a) Disparate process integration—Old was internal only, new includes external and power

   - Combine heretofore unrelated processes for more complete optimization, e.g.:

   – Desalination coupled with chemical manufacturing
   – Fertilizer production coupled with natural gas–based power generation, recovery, and distribution; gasification as a unifying tie process
   – Corn to chemicals
   – Power generation

An *unexpected benefit* of SE design is *localized power generation*. This comes at a most opportune time for the power industry that is near capacity and is already using consumer efficiency (energy star) improvements as a cost avoidance tool, removing the necessity of installing new power generation capacity. So, SE could not have come at a better time!

Additional SE benefits include avoidance of 20–30% transmission line losses, but will require teaming with power companies to coordinate interconnection to grid.

2.14 Teaching How to Design an Integrated Power and Chemical Production Facility: The SE Way

There are a few nontechnical components needing review if SE is to move forward, and a more practice-oriented preparation of undergraduate chemical engineers is one of them.

Included in Appendix C is an outline of the Capstone Engineering Process Design course, I taught for 6 years at the University of Illinois at Chicago (UIC), along with three examples of the fully integrated block flow diagrams that the students developed with their industrial mentors over three of those years. All were focused on a sustainable approach to ChE design engineering. I had a great deal of input from nearly a dozen of my design professional friends here in the Chicago area in creating the second-semester senior project. Our goal was to provide a real-world simulation of how-to for the students to crank on and get used to the idea of optimizing material and energy utilization. The interactive team development is meant to mirror industry and fits well with an emphasis on quality team building and cooperative design efforts. So why present this here?

1. It is difficult to get details like this from industry.
2. I was assisted by 6–10 fellow Chicago area design engineer colleagues in devising a complex, integrated power and chemical production facility—practice-based problems "ripped from the headlines."
3. Practice-based design teaching is of paramount importance if we are to inculcate our young men and women engineers of the future with SE design principles, a goal which my fellow practicing design engineers heartily agree with.
4. Our academic colleagues may find this useful in constructing their own senior capstone process design course.
5. Junior engineering colleagues may find this a useful guide for design as well as operations.
6. My senior engineering design colleagues will fill my mailbox with numerous suggestions for improvement!
7. Business managers can use this to "Get Smart" on, and get out in front of sustainability engineering benefits.

I have left this in syllabus outline form as it was developed as much as a design checklist as anything else that the graduates can take to their first job. The block diagram is the final result of interconnecting ten different components to make power as well as multiple merchant products. And the corn to chemicals example comes closest to sustainable as everything is made from corn, natures' own renewable resource. Another class project examined the conversion of Williston ND frack gas into chemical products which of course is not as sustainable in the pure sense of the word as natural gas is a finite resource. Still it too represents the manner in which disparate chemical production elements can be combined with plant power generation needed anyway in an extremely efficient manner.

2.15 Design Educational Standards: Challenges and Opportunities for SE

Another nontechnical SE component involves industrial practice standards. As already noted, academic preparation has continued on a more theoretical trajectory. The National Council of Examiners for Engineering and Surveying (NCEES) [5] is responsible for creating the PE licensing examination and its specification, which, in my opinion, encompasses the practice of ChE itself and forms, therefore, a practice competency standard. I have been involved with NCEES since 2007 and used this specification to inform and create my first-semester design syllabus covering basic engineering design methodologies and calculation procedures (Appendix B). So why present this here? Academia has become increasingly more focused on a more theoretical "Transport Phenomena" [6] approach, leading to an improved understanding of the *science* of chemical engineering. I became involved as the Accreditation Board for Engineering and Technology, Inc. (ABET) [7] requires practice-based instructors for the undergraduate senior capstone design courses, which we will examine later in Chap. 12 and Appendices B and C.

Noteworthy is the emerging academic study of *New Product Design* (Chap. 1, ref. [13] and Chap. 1, ref. [18]), as a preliminary first step to process engineering design. This coupled with the transport approach is arising from the generalized research-oriented graduate academic preparation and is applicable to SE. Industry will do well to take note as this will serve in identifying emerging technological trends applicable to SE, and by the way, it is also a great way to identify rising star students! In fact, tying academic research and development (R&D) back with industrial practice is really where the discipline of chemical engineering began, so this fits well with the focus on SE design advances in general.

Finally, the American Institute of Chemical Engineers (AIChE) provides a vibrant forum for industry/academia interface.

2.16 Only One Chance to Make a First Impression: Efficiency and the Bottom Line

Once process equipment is selected, purchased, installed and in operation, replacement for anything but end of equipment life, or breakdown is difficult to justify. Design professionals recognize the difference between theory and practice. In addition to meeting nameplate production throughput, economics demands minimizing the amount of energy consumed by all process equipment. If not done upfront, you will pay for poor performance over the entire project life! Chapter 4 examines the "efficiency of all things" and how careful design and selection of SE quality equipment helps in defining the long-term success of all projects. First things first, so next is a look at SE material and energy balances, sources, and sinks.

References

1. Luyben WL (2006) Distillation and control using aspen simulation. Wiley-AIChE, New York
2. Twigg MV (ed) (1996) Catalyst handbook, 2nd edn. Manson Publishing, London
3. Prausnitz J (1999). Molecular thermodynamics of fluid-phase equilibria. Wiley, New York
4. Jones DSJ, Pujado PR (eds) (2008) Handbook of petroleum processing. Springer, Cham
5. National Council of Examiners of Engineering and Surveying. www.ncees.org
6. Bird RB, Stewart WE, Lightfoot EN (1960) Transport phenomena, vol 108. Wiley, Hoboken, p 78C
7. Accreditation Board for Engineering and Technology, Inc. www.abet.org/accreditation/

Chapter 3
Material and Energy Sources and Sinks: More Power to You!

3.1 Seek Out and Combine Energy Sources and Sinks

This is not always possible within the restraints of single primary product classic process designs; so, it may require combining other, disparate product processes. For that matter, the easy and cheap way out regarding energy minimization is to avoid the expense of heat recovery, e.g., a cold stream against a hot one. Integrated power is the key.

3.2 The Btu Is the New Coin of the Realm

Or at least it should be. As described elsewhere, I worked on an energy evaluation project for a Walmart supplier. Turns out at the time, Walmart wisely required an energy analysis of all changes to things they bought from their suppliers. Fortunately, my evaluation showed an 85% reduction in Btu footprint for this plastic packaging material.

The US Energy Information Agency (EIA) [1] also goes to great pains to identify the efficiencies of various resources used to generate electricity, and this information is quite illuminating when evaluating alternative energy use. In particular, it is a fact-based tool that can help illustrate the potential savings by integrating power and chemical production.

After looking at this with respect to process design and improvement as well as expansion and improvement evaluations, it became apparent that at least an adjunct to pure $$ evaluation is needed. Just look at the wild swing in pricing for natural gas that arose from fracking. Over a few short years, natural gas pricing has fallen from $10 → $3 or so per million Btu. While this presents a difficult budget challenge, it is easy to normalize against energy consumption, at least for evaluation purposes.

J. P. Perl, *Sustainability Engineering for Enhanced Process Design and Manufacturing Profitability*, https://doi.org/10.1007/978-3-031-52363-2_3

The temptation to invest purely on the basis of short-term economic conditions is great and common. In particular, where large energy pricing variations can happen over relatively short periods of time, some other basis must be used. I guess the entire agricultural futures market provides a hedge against that kind of variability in that market segment. Dual or multi-fuel use can serve as a hedge against large swings in energy pricing.

BIG IDEA—We may need a sort of SE hedge fund to help normalize the risk of using high-value SE resources, both energy and materials, that are subject to volatile international economics.

3.3 New Product/Process Design or Process Changes

In addition to new process design, all contemplated process changes should also have both materials and energy footprint analysis for SE to operate. Of course, we cannot invest in losing propositions, but the range in energy consumption can also provide inputs for confidence level analysis. It can also highlight the need to identify more stable, sustainable supplies. This provides another way of looking at things beyond "That's the way we have always done it." Shine a light on it so all in the quality circle can work on it, or at least be aware of it. President Franklin Roosevelt gave pre-WWII American industry a request to produce more than ten times the rate than ever done before [2]. He was told it was impossible, but refused to believe it and said American industry would find a way … and they did by out-producing even FDR's wildest dreams, and in the process established new benchmarks for automated production and utilization of energy and materials. As a side note, shortly after WWII, Dr. W. Edwards Deming transplanted his total quality management (TQM) system into Japan during the American military occupation [3]. The Japanese absorbed and adopted this rapidly, and by the 1970s were producing all manner of manufactured goods of superior quality to, and cheaper than the USA. The post-WWII US economy enjoyed little global competition, so we were slower to adopt any changes. (We did make the best tanks, ships, and warplanes, but few Americans had room for them in the driveway!) We soon learned that as Phil Crosby said, "Quality is free" [4]. The Japanese raised the bar for quality-based manufacturing worldwide. And now it is time to up the ante with SE principles.

3.4 Material and Energy Balances: Nothing Has Changed!

I am not sure when I came to this as a young engineer, but I do view the system of material and energy balances used by chemical engineers as a point of departure for nearly everything we do. As such, it is a sort of double-entry bookkeeping accounting system for chemical engineers. The two go hand in glove although we normally perform a material balance as a first step with the energy balance as a double check. Of course, both are needed anyway to evaluate process economics. (So working in my cousin's accounting firm came in handy after all!)

Mechanical and HVAC engineers will find this accounting method similar to evaluating HVAC and compressor power requirements when amperage and voltage are known, using the conservation of energy principle. The energy consumption of a compressor is multiplied by the mass flow of refrigerant to yield actual power demand. This is compared to vendor performance tables to verify operation, commissioning of new installations, and troubleshoot existing ones. By definition, the vendor tables account for their specific equipment efficiency factors, but the bottom line is still how many electrical watts will it take to move a given thermal load in Btu/h, expressed as the energy efficiency ratio (EER), where EER = 3.41214 × COP and COP = Btu Delivered/Btu Used. So, EER tracks COP and the vendor tables represent the variability of COP with thermal load. It is this variability that always needs to be factored into any design where mass flowrate fluctuations are expected, and in particular, where disparate processes are tied together.

Identifying and matching energy sources and sinks are fundamental to maximizing utilization and minimizing cost. During the process design phase, a simple picture, or block flow diagram (BFD), of the desired process is first drawn. A material balance is then made on all streams and labeled on the BFD. Next, a process flow diagram (PFD) is started that includes an energy balance. Information and data on the PFD linking the material and energy required to heat, cool, separate, or react is then tabulated.

As an example, if water evaporation is required, the material into the evaporator might be 100 #/h of liquid water at 212 °F and the output might be water vapor at say 14.7 psia saturated. Again, our double-entry bookkeeping approach comes into play as we note that 100 pounds of liquid water disappears and 100 pounds of water vapor (steam) appears. Concurrent to this is the loss (sink) of 970,000 Btu/h from the external energy source to the steam. This double-entry bookkeeping system is the bread and butter and the basis of chemical process design and troubleshooting, and uses the conservation of mass and energy laws as a cross-check of the books from both ends. Deviation from this balance could be due to process inefficiency, design, measurement error, or in the case of materials, an unwanted or accidental loss/release to the environment, or workplace exposure.

This in and of itself is insufficient to support SE design. The preceding was an example of an energy sink. For an application such as this, steam is often used as a heating element to cause evaporation and is therefore referred to as a source. So, as a first step, the design engineer simply shows these streams on the BFD and makes sure that all materials are balanced, and that the energy required for heating, cooling, or reacting is accounted for. A good first step, necessary but insufficient for SE.

3.5 Electricity or Motive Power from Steam

Energy quality is a key concept for SE in the CPI. The old-time steam engineers defined steam quality as 100—the mass percent of liquid water droplets in the stream of liquid and vapor. Liquid either ruins compressors or takes away work/pound of steam from condensers. So, a high-quality steam is actually low in liquid water, with 100% representing pure vapor. Just check your American Society of Mechanical Engineers (ASME) steam tables and Mollier diagram, it still shows up.

3.6 Energy in General

There is an additional energy quality definition that I find useful in SE and that is based on the Carnot or other cycle, as well as basic process thermodynamics. Carnot: $[(T_{HIGH} - T_{LOW})/T_{HIGH}] \times 100$ = classic thermodynamic efficiency where temperatures are absolute, either Kelvin or Rankine. For such a process to be efficient, a very high delta T is required. When generating electricity in a turbine, once the high temperature and pressure (T&P) steam is reduced through expansion, in order to capture its energy for electricity generation, what remains is a vast material quantity of "low-quality heat." In combined heat and power (CHP) plants, heat recovery steam generators (HRSG), capture this within the electricity generation process through various expansion stages. But we want more for the CPI!

3.7 Integrated Power and Chemical Production

This is a major, pivotal element for SE design. The CPI is best positioned to take maximum advantage here. This is huge and has broad applicability across a number of industry segments as we will see in the following sections.

3.8 High Volume—Low-Quality Heat Recovery in the CPI and Elsewhere

3.8.1 Existing

This allows even warm liquid, e.g., at 100 °F to be used to heat a cold feed stream. Here, low-quality energy has little to no cyclic power generation capability. Through the use of heat exchangers, a chemical process can take low-quality hot liquid to warm up a cold process feed stream. The equation for this is Q = mass times specific heat times the temperature change in the low-quality stream, or $M \times Cp \times \Delta T$. This is then used to size the heat exchange equipment via the design equation: Q = Overall heat transfer coefficient times the heat transfer area times the log mean temperature difference (LMTD), taken this time across the hot and cold streams, or $Q = U\text{oa} \times A \times$ Log Mean Temperature Difference (LMTD). From this, one can solve for the size (area) of the desired heat exchanger if the terminal temperatures are chosen, or for a fixed area, determine those same temperatures. With size comes the cost to purchase and operate. Watch out for poor overall heat transfer coefficient (Uoa) which may be a fact you have to live with or signal low-efficiency equipment which should be addressed upfront in SE design or live with long-term cost consequences.

3.8.2 *New SE*

As discussed elsewhere, this concept is not limited to power but also to the integration of disparate chemical process, exothermic and endothermic, by nature, but with insufficient energy to drive Carnot engines. My mechanical engineering (MMAE) buddies will certainly start looking to team with chemical engineering (ChE)!

The careful matching of disparate processes is called for to optimize SE designs. The concept of exergy is another way of evaluating the energy actually available within process streams [5].

Once an understanding of the basic process (design basis) is clear, the ESOH and materials sourcing evaluation must be completed to close the SE loop. Worthy of repetition is the need for early, preliminary evaluation of ESOH to assure permitability and safety prior to expending considerable engineering efforts. Permitability ≡ The ability to receive a permit to operate in compliance with regulations.

Two integrated power examples are shown in Figs. 8.5 and 8.6, and described in greater detail in Appendix C. A good basic SE without power but integrating heat and other materials is shown in Chap. 8, Sect. 8.4, Reforming Natural Gas to Chemicals and Clean Fuels.

3.9 Renewable and Other Material Sourcing

Renewable resources include crops, wind, and solar power. Oil and natural gas are not, strictly speaking, renewable. Fusion nuclear might someday participate if the technical difficulties can be overcome; ocean thermal energy, wave action, hydropower, biomass geothermal, biofuels, and H_2 from renewables are sort of next in line, perhaps serving as bridges to the future. Nuclear breeder reactor, converting bombs to power, might someday help the equation as well. But really, we basically have the sun and photosynthesis and the Krebs cycle of life as permanent renewable resources as the source of all life on the planet.

Gasification is a well-developed and incredibly versatile technology. With the proper front-end design on materials handling and preparation, just about anything organic can be safely and economically turned into CO and H_2, basic building blocks of chemical reaction engineering; that is, if properly integrated into process and energy production.

3.10 Gasification: New SE Design Tool for Material and Energy Integration

Some examples of these are presented in another section, but let us take a look at some established chemistries that are ripe for SE application (Chap. 1, Sect. 1.8).

3.11 Gasification of Various Organic Resources

1. Gasification of natural gas
2. Gasification of municipal solid waste (MSW)
3. Gasification of coal
4. Gasification of waste hydrocarbon products, wood, paper, etc.
5. CO_2 chemical processing utilization … not sequestration!

3.12 Gasification Chemistries and Product Pathways

With the advent of enhanced natural gas production via fracking, an enormous new supply of a versatile feedstock, i.e., methane and other related compounds, has suddenly made itself available. Gasification will eventually earn its place in the pantheon of SE technology as it is the consummate reformer of all things organic, including waste by-products. In fact, as it serves to rescue otherwise stranded, or useless materials, it may even be considered a material itself!

(Note: Higman, et al Reference 8 in Chap. 1 is an excellent Gasification resource, and I have reproduced only a few equations here.)

The principal reactions include:

Steam Methane Reforming Reaction:

$$CH_4 + H_2O \rightleftarrows CO + 3H_2$$

Water GasShift Reaction:

$$CO + H_2O \rightleftarrows CO_2 + H_2$$

Hydrogen is a very important feedstock for the production of ammonia. CO_2 is important for the production of Urea-Ammonium Nitrate (UAN). These fertilizer components are ubiquitous.

3.13 CO_2 as a Feedstock

This is a holy grail of SE as the primary recycle pathway for vapor-phase CO_2 to carbon product. A few examples follow:

LanzaTech Process (detailed in Chap. 8). This process takes place in a fermentation bioreactor in the presence of proprietary microbes.

$$CO_2 + 3CH_4 \rightleftarrows 2(C_2H_5OH)$$

Black and Veach Process (Fig. 8.4):

$$CO_2 + Syngas \rightarrow Methanol \rightarrow Gasoline$$

Urea produced from Gasifier ammonia:

$$2NH_3 + CO_2 \rightleftarrows NH_2CONH_2 + H_2O$$

Urea Ammonium Nitrate (UAN) via Stamicarbon's CO_2 Stripping Process—Stamicarbon [6]. The industrial production of ammonium nitrate (AN) from ammonia and nitric acid:

$$HNO_{3+} NH_3 \rightarrow NH_4 NO_3$$

MIDREX [7] Chemistries—Iron Ore oxidation for mini-mill feedstock from gasified natural gas:

$$2Fe_2O_3 + 2CO + 2H_2 \rightleftarrows 4Fe + 3H_2O + 3CO_2$$

Here, H_2 and CO from the gasifier are used in the direct reduction of iron (DRI) in the Midrex process. This is most effective when located adjacent to steel production mini-mill so the solid iron product can be fed hot to the mill. Here is where the SE integrated power and process concept really comes into play. The Iron would otherwise need to be cooled and reheated, a very inefficient process to begin with. We'll look at an example in Chap. 8.

3.14 Material Sourcing Summary

An important component of the sustainability engineering design glue holding things together is the combination of gasification, electric generation, natural gas cleanup, and total heat recovery. An additional plus is the incorporation, not mere sequestration, of CO_2 as a feedstock. These reactions are examined again in Chap. 8.

3.15 Energy Sourcing

The key to successful and optimal SE is the use of renewable energy and material resources.

3.15.1 *Finite Nonrenewable Resources*

Energy from natural gas, oil, coal, electricity from hydro, fossil fuels, and nuclear are not, strictly speaking, renewable.

3.15.2 *Renewables*

Hydroelectric, geothermal, solar thermal, photovoltaics, wind, and biomass are examples. From the EIA 2011 Annual Energy Report, Table 1.3 reproduced on the next page, renewables prior to 2001 accounted for a fluctuating average of $7.2 \pm 2\%$. After 2001, this has been increasing from 5.4% in 2001 to 9.3% in 2011.

3.16 Common Often Unused (Stranded, Wasted) Energy Sources: A Bridge to SE

Many of the following waste heat examples can be converted to reuse, thereby becoming semi-renewables themselves. While not infinitely renewable, they can greatly multiply energy availability and serve as a bridge into SE mode.

- Pumping loops—A large amount of energy is consumed by pumping fluids around a plant so overall pumping design layout for energy minimization while maintaining operability, functionality, and maintainability must be done. The Department of Energy (DOE) estimates 15–30% energy loss in fluid-fluid plants [8].

- Compressed air—This is needed to operate control valves and blanket systems (N2). Leaks are common leading to huge long-term energy loss and should be avoided if electric or other actuators can be used instead. Electrical control of systems also can be more precise.
- Distillation—Heat energy is the driving force for this ubiquitous separation process, and much energy is lost in condensers and coolers that can be recovered by use of economizers, mechanical vapor recompression (MVR), dividing wall (7.14) and placement amongst low-quality heat sinks, e.g., reactors and mixers, need to be considered here. Refineries already do this well, but not always throughout the CPI.
- Evaporation—A very common process, particularly in the food industry, often representing totally lost latent heat of water. Industry-specific regulations, e.g., FDA will proscribe contacting equipment limits, but an enormous amount of energy could be captured at least by using HRSG to make electricity in food evaporation processes or by combining with other processes where contamination will not be an issue.

✈And always be on the lookout for government and/or utility incentives for co-generation.✈.

3.17 Onsite Integrated Electricity Generation: An SE Mainstay

1. Reduces Transmission Line Loss = Efficiency Booster
2. Provides Electric Utility Cost Avoidance = Economic Bottom Line Improvement
3. Reduce Need for Additional Power Plants

   (a) Saves fuel and reduces carbon footprint

The following two Tables 3.1 and 3.2, are excerpted from EIA.Gov. I have left the original labels 1.3 and 10.1, respectively to maintain EIA reference. The first (EIA table 1.3) shows primary energy consumption from 1949 to 2011. This is followed by (EIA table 10.1) an expansion of the renewable energy values used for that period. This is part of their 2011 Annual Report [9], and is an excellent resource for determining baseline energy availability in physically realizable quantity as relates to US energy production and consumption. This also gives some idea of alternate energy sources for available SE applications. As can be seen, we have a long way to go if renewable energy is to contribute a major portion of our consumption, providing both challenges and opportunities. The interested reader is directed to https://www.eia.gov/totalenergy/data/annual/pdf/aer.pdf for this original as well as the new monthly format reports.

Integrated energy and chemical production as well as improved efficiency and recycling will provide the largest SE impact going forward. The opportunity to save money while improving the SE footprint is enormous. And once again, reduction in

Table 3.1 Primary energy consumption estimates by source, USA, 1949–2011

Year	Fossil Fuels					Nuclear Electric Power	Renewable Energy [1]					Electricity Net Imports [3]	Total
							Noncombustible [2]						
	Coal	Coal Coke Net Imports [3]	Natural Gas [4]	Petroleum [5]	Total		Captured Energy [6]	Adjustment for Fossil Fuel Equivalence [6]	Total [6,7]	Biomass [7]	Total		
1949	11.981	−0.007	5.145	11.883	29.002	0.000	0.323	1.101	1.425	1.549	2.974	0.005	31.982
1950	12.347	.001	5.968	13.315	31.632	.000	.344	1.071	1.415	1.562	2.978	.006	34.616
1955	11.167	−.010	8.998	17.255	37.410	.000	.397	.963	1.360	1.424	2.784	.014	40.208
1960	9.838	−.006	12.385	19.919	42.137	.006	.510	1.098	1.608	1.320	2.928	.015	45.086
1965	11.581	−.018	15.769	23.246	50.577	.043	.673	1.388	2.061	1.335	3.396	(s)	54.015
1970	12.265	−.058	21.795	29.521	63.522	.239	.858	1.781	2.639	1.431	4.070	.007	67.838
1975	12.663	.014	19.948	32.732	65.357	1.900	1.045	2.143	3.188	1.499	4.687	.021	71.965
1976	13.584	(s)	20.345	35.178	69.107	2.111	.991	2.022	3.014	1.713	4.727	.029	75.975
1977	13.922	.015	19.931	37.124	70.991	2.702	.775	1.595	2.371	1.838	4.209	.059	77.961
1978	13.766	.125	20.000	37.963	71.854	3.024	.977	1.990	2.968	2.038	5.005	.067	79.950
1979	15.040	.063	20.666	37.122	72.891	2.776	.979	1.992	2.971	2.152	5.123	.069	80.859
1980	15.423	−.035	20.235	34.205	69.828	2.739	.970	1.983	2.953	2.476	5.428	.071	78.067
1981	15.908	−.016	19.747	31.932	67.571	3.008	.920	1.898	2.817	2.596	5.414	.113	76.106
1982	15.322	−.022	18.356	30.232	63.888	3.131	1.082	2.234	3.316	2.663	5.980	.100	73.099
1983	15.894	−.016	17.221	30.052	63.152	3.203	1.165	2.426	3.591	2.904	6.496	.121	72.971
1984	17.071	−.011	18.394	31.053	66.506	3.553	1.133	2.334	3.467	2.971	6.438	.135	76.632
1985	17.478	−.013	17.703	30.925	66.093	4.076	1.002	2.066	3.068	3.016	6.084	.140	76.392
1986	17.260	−.017	16.591	32.198	66.033	4.380	1.038	2.141	3.179	2.932	6.111	.122	76.647
1987	18.008	.009	17.640	32.864	68.521	4.754	.900	1.847	2.747	2.875	5.622	.158	79.054
1988	18.846	.040	18.448	34.223	71.557	5.587	.807	1.634	2.441	3.016	5.457	.108	82.709
1989	19.070	.030	19.602	34.209	72.911	5.602	1.048	2.028	3.076	3.159	6.235	.037	84.786

1990	19.173	.005	19.603	33.552	72.332	6.104	1.128	2.177	3.306	2.735	6.041	.008	84.485
1991	18.992	.010	20.033	32.846	71.880	6.422	1.121	2.166	3.287	2.782	6.069	.067	84.438
1992	19.122	.035	20.714	33.525	73.396	6.479	1.001	1.889	2.890	2.932	5.821	.087	85.783
1993	19.835	.027	21.229	33.745	74.836	6.410	1.100	2.074	3.174	2.908	6.083	.095	87.424
1994	19.909	.058	21.728	34.561	76.256	6.694	1.030	1.930	2.961	3.028	5.988	.153	89.091
1995	20.089	.061	22.671	34.438	77.259	7.075	1.197	2.262	3.459	3.101	6.560	.134	91.029
1996	21.002	.023	23.085	35.675	79.785	7.087	1.326	2.530	3.857	3.157	7.014	.137	94.022
1997	21.445	.046	23.223	36.159	80.873	6.597	1.360	2.550	3.910	3.105	7.016	.116	94.602
1998	21.656	.067	22.830	36.816	81.369	7.068	1.247	2.318	3.565	2.927	6.493	.088	95.018
1999	21.623	.058	22.909	37.838	82.427	7.610	1.240	2.312	3.552	2.963	6.516	.099	96.652
2000	22.580	.065	23.824	38.262	84.731	7.862	1.090	2.008	3.098	3.008	6.106	.115	R98.814
2001	21.914	.029	22.773	38.186	82.902	8.029	.893	1.647	2.540	2.622	5.163	.075	96.168
2002	21.904	.061	R23.510	38.224	R83.699	8.145	1.070	1.959	3.029	2.701	5.729	.072	R97.645
2003	22.321	.051	22.831	38.811	84.014	7.959	1.114	2.062	3.176	2.807	5.983	.022	97.978
2004	22.466	.138	R22.923	40.292	R85.819	8.222	1.103	1.969	3.073	3.010	6.082	.039	R100.162
2005	22.797	.044	R22.565	40.388	R85.794	8.161	1.127	1.998	3.125	R3.117	6.242	.085	R100.282
2006	22.447	.061	R22.239	39.955	R84.702	8.215	1.229	2.153	3.382	R3.267	R6.649	.063	R99.629
2007	22.749	.025	R23.663	39.774	R86.211	8.455	1.125	1.924	3.048	R3.474	R6.523	.107	R101.296

(continued)

Table 3.1 (continued)

Year	Fossil Fuels					Nuclear Electric Power	Renewable Energy [1]					Electricity Net Imports [3]	Total
							Noncombustible [2]						
	Coal	Coal Coke Net Imports [3]	Natural Gas [4]	Petroleum [5]	Total		Captured Energy [6]	Adjustment for Fossil Fuel Equivalence [6]	Total [6,7]	Biomass [7]	Total		
2008	22.385	.041	R23.843	37.280	R83.549	8.427	1.238	2.099	3.338	R3.849	R7.186	.112	R99.275
2009	19.692	–.024	R23.416	35.403	R78.488	8.356	1.382	2.306	3.688	R3.912	R7.600	.116	R94.559
2010	R20.850	–.006	R24.256	R36.010	R81.109	R8.434	R1.440	R2.355	R3.796	R4.294	R8.090	R.089	R97.722
2011P	19.643	.011	24.843	35.283	79.779	8.259	1.785	2.939	4.724	4.411	9.135	.127	97.301

Authors note: This appeared as Table 1.3 in the original EIA report-jpp

[1]Most data are estimates. See Note, "Renewable Energy Production and Consumption," at end of Section 10

[2]Conventional hydroelectric power, geothermal, solar thermal, photovoltaic, and wind. See Note 1, "Noncombustible Renewable Energy," at end of section

[3]Net imports equal imports minus exports. A minus sign indicates exports are greater than imports

[4]Natural gas only; excludes supplemental gaseous fuels. See Note 1, "Supplemental Gaseous Fuels," at end of Section 6

[5]Petroleum products supplied, including natural gas plant liquids and crude oil burned as fuel. Does not include biofuels that have been blended with petroleum—biofuels are included in "Biomass." For petroleum, product supplied is used as an approximation of petroleum consumption. See Note 1, "Petroleum Products Supplied and Petroleum Consumption," at end of Section 5

[6]See Note 1, "Noncombustible Renewable Energy," at end of section

[7]See Table 10.1 for a breakdown of individual sources

R=Revised. P=Preliminary. (s)=Less than 0.0005 and greater than –0.0005 quadrillion Btu

Notes: • See "Primary Energy Consumption" in Glossary. • See Table E1 for estimated energy consumption for 1635–1945. • See Note 3, "Electricity Imports and Exports," at end of Section 8. • Totals may not equal sum of components due to independent rounding

Web Pages: • See http://www.eia.gov/totalenergy/data/monthly/#summary for updated monthly and annual data. • See http://www.eia.gov/totalenergy/data/annual/#summary for all annual data beginning in 1949

Sources: Tables 5.12, 6.1, 7.1, 7.8, 8.1, 8.2a, 10.1, 10.3, A4, A5, and A6

Table 3.2 Renewable energy production and consumption, by primary energy source USA, 1949–2011

Year	Production [1]			Consumption								
	Biomass		Total Renewable Energy [4]	Hydroelectric Power [5]	Geothermal[6]	Solar/ PV [7]	Wind [8]	Biomass				Total Renewable Energy
	Biofuels [2]	Total [3]						Wood [9]	Waste [10]	Biofuels [11]	Total	
1949	NA	1549	2974	1425	NA	NA	NA	1549	NA	NA	1549	2974
1950	NA	1562	2978	1415	NA	NA	NA	1562	NA	NA	1562	2978
1955	NA	1424	2784	1360	NA	NA	NA	1424	NA	NA	1424	2784
1960	NA	1320	2928	1608	(s)	NA	NA	1320	NA	NA	1320	2928
1965	NA	1335	3396	2059	2	NA	NA	1335	NA	NA	1335	3396
1970	NA	1431	4070	2634	6	NA	NA	1429	2	NA	1431	4070
1975	NA	1499	4687	3155	34	NA	NA	1497	2	NA	1499	4687
1976	NA	1713	4727	2976	38	NA	NA	1711	2	NA	1713	4727
1977	NA	1838	4209	2333	37	NA	NA	1837	2	NA	1838	4209
1978	NA	2038	5005	2937	31	NA	NA	2036	1	NA	2038	5005
1979	NA	2152	5123	2931	40	NA	NA	2150	2	NA	2152	5123
1980	NA	2476	5428	2900	53	NA	NA	2474	2	NA	2476	5428
1981	13	2596	5414	2758	59	NA	NA	2496	88	13	2596	5414
1982	34	2663	5980	3266	51	NA	NA	2510	119	34	2663	5980
1983	63	2904	6496	3527	64	NA	(s)	2684	157	63	2904	6496
1984	77	2971	6438	3386	81	(s)	(s)	2686	208	77	2971	6438
1985	93	3016	6084	2970	97	(s)	(s)	2687	236	93	3016	6084
1986	107	2932	6111	3071	108	(s)	(s)	2562	263	107	2932	6111
1987	123	2875	5622	2635	112	(s)	(s)	2463	289	123	2875	5622
1988	124	3016	5457	2334	106	(s)	(s)	2577	315	124	3016	5457
1989	125	3159	6235	2837	162	55	22	2680	354	125	3159	6235
1990	111	2735	6041	3046	171	59	29	2216	408	111	2735	6041

(continued)

Table 3.2 (continued)

Year	Production [1]			Consumption									
	Biomass		Total Renewable Energy [4]	Hydroelectric Power [5]	Geothermal[6]	Solar/ PV [7]	Wind [8]	Biomass				Total Renewable Energy
	Biofuels [2]	Total [3]						Wood [9]	Waste [10]	Biofuels [11]	Total	
1991	128	2782	6069	3016	178	62	31	2214	440	128	2782	6069
1992	145	2932	5821	2617	179	64	30	2313	473	145	2932	5821
1993	169	2908	6083	2892	186	66	31	2260	479	169	2908	6083
1994	188	3028	5988	2683	173	68	36	2324	515	188	3028	5988
1995	198	3099	6558	3205	152	69	33	2370	531	200	3101	6560
1996	141	3155	7012	3590	163	70	33	2437	577	143	3157	7014
1997	186	3108	7018	3640	167	70	34	2371	551	184	3105	7016
1998	202	2929	6494	3297	168	69	31	2184	542	201	2927	6493
1999	211	2965	6517	3268	171	68	46	2214	540	209	2963	6516
2000	233	3006	6104	2811	164	R66	57	2262	511	236	3008	6106
2001	254	2624	5164	2242	164	64	70	2006	364	253	2622	5163
2002	308	2705	5734	2689	171	63	105	1995	402	303	2701	5729
2003	402	2805	5982	2825	175	62	115	2002	401	404	2807	5983
2004	487	2998	6070	2690	178	63	142	2121	389	499	3010	6082
2005	564	3104	6229	2703	181	63	178	R2,137	403	577	R3,117	6242
2006	720	R3,216	R6,599	2869	181	68	264	R2,099	397	771	R3,267	R6,649
2007	978	R3,461	R6,509	2446	186	76	341	R2,070	413	991	R3,474	R6,523
2008	1387	R3,864	R7,202	2511	192	89	546	R2,040	436	1372	R3,849	R7,186
2009	R1,584	R3,928	R7,616	2669	200	98	721	R1,891	R453	R1,568	R3,912	R7,600
2010	R1,884	R4,341	R8,136	R2,539	R208	R126	R923	R1,988	R469	R1,837	R4,294	R8,090
2011P	2047	4511	9236	3171	226	158	1168	1987	477	1947	4411	9135

Authors Note: This appeared as Table 10.1 in the original EIA report

[1]Production equals consumption for all renewable energy sources except biofuels

[2]Total biomass inputs to the production of fuel ethanol and biodiesel

[3]Wood and wood-derived fuels, biomass waste, and total biomass inputs to the production of fuel ethanol and biodiesel

[4]Hydroelectric power, geothermal, solar thermal/photovoltaic, wind, and biomass

[5]Conventional hydroelectricity net generation (converted to Btu using the fossil-fuels heat rate—see Table A6)

[6]Geothermal electricity net generation (converted to Btu using the fossil-fuels heat rate—see Table A6), and geothermal heat pump and direct use energy

[7]Solar thermal and photovoltaic (PV) electricity net generation (converted to Btu using the fossil-fuels heat rate—see Table A6), and solar thermal direct use energy

[8]Wind electricity net generation (converted to Btu using the fossil-fuels heat rate—see Table A6)

[9]Wood and wood-derived fuels

[10]Municipal solid waste from biogenic sources, landfill gas, sludge waste, agricultural byproducts, and other biomass. Through 2000, also includes non-renewable waste (municipal solid waste from non-biogenic sources, and tire-derived fuels)

[11]Fuel ethanol (minus denaturant) and biodiesel consumption, plus losses and co-products from the production of fuel ethanol and biodiesel

R=Revised. P=Preliminary. NA=Not available. (s)=Less than 0.5 trillion Btu

Notes: • Most data for the residential, commercial, industrial, and transportation sectors are estimates. See notes and sources for Tables 10.2a and 10.2b. • See Tables 8.2a–8.2d and 8.3a–8.3c for electricity net generation and useful thermal output from renewable energy sources; Tables 8.4a–8.4c, 8.5a–8.5d, 8.6a–8.6c, and 8.7a–8.7c for renewable energy consumption for electricity generation and useful thermal output; and Tables 8.11a–8.11d for renewable energy electric net summer capacity. • See Note, "Renewable Energy Production and Consumption," at end of section. • See Table E1 for estimated renewable energy consumption for 1635–1945. • Totals may not equal sum of components due to independent rounding

Web Pages: • See http://www.eia.gov/totalenergy/data/monthly/#renewable for updated monthly and annual data. • See http://www.eia.gov/totalenergy/data/annual/#renewable for all annual data beginning in 1949. • See http://www.eia.gov/renewable/ for related information

Sources: **Biofuels:** Tables 10.3 and 10.4. **All Other Data:** Tables 10.2a–10.2c

energy consumption through SE design and production methodologies also leads to reduced carbon footprint and attendant impacts. As US manufacturing becomes more SE-based, it actually becomes more economically competitive as well.

3.18 Common Often Unused (Wasted) Material Sources

Almost all of the following are candidates for gasification:

- Municipal solid waste (MSW), aka garbage, is a rich source of carbon, metals, and glass for recycling or repurposing
- Corn husks, bagasse from sugar cane, and similar materials for gasification feedstock
- Foresting operations—instead of controlled burns—cut down and use in bioprocesses
- Environmental remediation, e.g., organic and inorganic containing soils and spent solvents
- Production waste materials not consumed, but recoverable as energy by reformatting
- Nonrecyclable plastics that could be thermally reformed to other chemicals

3.19 Material and Energy Integration Approaches: A New Approach for SE

Combined heat and power typically refers to the capture of steam boiler waste heat for power use but often still strands low-quality heat source. However, through SE design practice, the CPI can use most of the low-quality stranded heat. Here, low-quality heat is defined as unusable in Carnot engines and for the case of steam used for motive or expansion purpose, wet steam is considered lower quality steam. The optimal use of low-quality heat (LQH) arises from combining processes not commonly found together, e.g., gasification and the Midrex iron ore reduction process. For example, using a hot solid waste stream exiting from one process directly as feed into thc next process, without the need to cool and reheat, can lead to nearly 100% energy recovery, a feat not possible with thermodynamic engines (Carnot and other cycles). An example can be $Fe_2O_3 \rightarrow Fe$, reduction of iron ore in combined heat and power applications (covered later in the SE example section).

The arena of integrated power and chemical production is perhaps the most exciting opportunity for SE design and is now starting to find applications. Sulfuric acid plants have long co-located next to refineries as sulfur (Ch2:4), a common petroleum feed contaminant is often removed in copious quantities from crude oil feed stocks to prevent downstream catalyst poisoning and equipment corrosion. We

Table 3.3 Resources for classic (present) and SE design

	Present use	Future SE use
Nonrenewables		
Crude oil	Refinery feedstock, gasoline, etc.	Plastics and petrochemicals
Natural gas	Heating, cooking, boiler, turbine	Electric and chemical production
Metals	Electrics, construction	Electronics, construction
Renewables		
Photovoltaics	Sensors, small electric generation	Sensors, large-scale generation
Solar heating	Water, power	Same, larger scale
Trees	Construction, furniture, paper, fuel	Construction, furniture, paper
Algae to oil	Under development	Motive fuels
CO_2	With gasifier—Fertilizer	With gasifier—Fertilizer
MSW	Some CH_4, wasted land use	Mine metals and glass, gasify rest
Hazardous waste	Incineration—Total loss	Gasify

will look at some of these examples later. In fact, the modern oil refinery is a great place to start to observe creative methods of energy and material conservation (Ch2:4).

3.20 SE Classification of Resources for Production

With reference to Table 3.3 above:

1. Renewable
   (a) Solar heating, algae when developed, trees when properly husbanded, photovoltaics when better developed
   (b) CO_2as a feedstock
   (c) Crop production
   (d) Forest production
   (e) Nuclear—fusion
   (f) MSW
2. "Bridging" transition to SE
   (a) Frack gas as a source of electric generation, chemical production, and CO_2 sequestration when used in fertilizer production
   (b) Coal and tar sands when gasified under mild conditions that control emissions to create the same syngas as frack gas (mining activity must be SE)

   (c) Municipal solid wastes after plastic, glass, and metal recyclables removed; gasify to create power, fuel, and chemical production feedstock; this class is an irreducible by-product of human activity

3. Nonrenewable

   (a) Oil—transition from transportation to use as a high-value-added petrochemical feedstock
   (b) Natural gas and metals
   (c) Nuclear, fission

3.21 Common Commercial Recyclables and Handling

1. Metals, paper, plastics, glass—the usual suspects
2. Electronics

   (a) Computers, TV, etc.

   - Metals, precious metal, and organics recovery
   - Gasify the rest!

3. Construction debris

   (a) Concrete to aggregate
   (b) Wood to paper
   (c) Metals to augment mining

4. Water—Goal is zero discharge and total reuse

   (a) Human wastewater
   (b) Industrial wastewater
   (c) Agricultural
   (d) Ocean desalination + integrated power and chemical production

5. Chemical

   (a) Waste oil and lubricants
   (b) solvents
   (c) process waste chemicals usable elsewhere

6. Enhanced municipal waste segregation

   (a) Gasification or pyrolysis of municipal solid waste (MSF)

3.22 SE Design: Bridges to the Future Needing Continued Cost Efficiency Improvement

1. Photosynthetics
2. Photovoltaics
3. Minimal nuclear waste and develop nuclear fusion reactors
4. Reprocess bomb material—swords to plowshares

3.23 Artificial Leaf Harnesses Sunlight for Efficient Fuel Production [10]

> *Generating and storing renewable energy, such as solar or wind power, is a key barrier to a clean-energy economy. When the Joint Center for Artificial Photosynthesis (JCAP) was established at Caltech and its partnering institutions in 2010, the U.S. Department of Energy (DOE) Energy Innovation Hub had one main goal: a cost-effective method of producing fuels using only sunlight, water, and carbon dioxide, mimicking the natural process of photosynthesis in plants and storing energy in the form of chemical fuels for use on demand. Over the past five years, researchers at JCAP have made major advances toward this goal, and they now report the development of the first complete, efficient, safe, integrated solar-driven system for splitting water to create hydrogen fuels.*

✈This one will compete for the sun with algal oil and perhaps photovoltaics. Let's see who wins first.✈.

3.24 Geothermal Energy

This sometimes refers to thermally active regions close to the surface, e.g., hot springs and geysers, more often to using the earth as a partner in energy conservation. Residential, commercial, and industrial applications are in great use already.

3.24.1 As Source or Sink for Low to Medium Thermal Loads: Residential/Light Commercial

Go down about 50–150 ft. below the earth's surface, and the temperature is relatively stable. Typical temperature of 55 °F is not uncommon. For small-scale thermal loading, using the earth as a massive heat sink/source is ideal for the improvement of heat pump heating and air conditioning systems. In the winter, the heat pump coefficient of performance can exceed 5.0, meaning 80% reduction in electric heating costs. Where natural gas is available, however, a condensing furnace can attain

95% heat recovery, so the analysis always must be done first on a Btu footprint basis and then of course on a dollar basis.

During the cooling season, savings on the order of 40% can be had due to the low, consistent, subsurface temperature of condensation, e.g., 95 °F air versus 50 °F subsurface and therefore greatly improved heat removal operation.

According to the excerpt from US DOE (https://www.energy.gov/energysaver/geothermal-heat-pumps):

Geothermal heat pumps (GHPs), sometimes referred to as GeoExchange, earth-coupled, ground-source, or water-source heat pumps, have been in use since the late 1940s. They use the relatively constant temperature of the earth as the exchange medium instead of the outside air temperature.

Although many parts of the country experience seasonal temperature extremes—from scorching heat in the summer to sub-zero cold in the winter—a few feet below the earth's surface the ground remains at a relatively constant temperature. Depending on latitude, ground temperatures range from 45 °F (7 °C) to 75 °F (21 °C). Like a cave, this ground temperature is warmer than the air above it during the winter and cooler than the air in the summer. The GHP takes advantage of these more favorable temperatures to become high efficient by exchanging heat with the earth through a ground heat exchanger.

As with any heat pump, geothermal and water-source heat pumps are able to heat, cool, and, if so equipped, supply the house with hot water. Some models of geothermal systems are available with two-speed compressors and variable fans for more comfort and energy savings. Relative to air-source heat pumps, they are quieter, last longer, need little maintenance, and do not depend on the temperature of the outside air.

A dual-source heat pump combines an air-source heat pump with a geothermal heat pump. These appliances combine the best of both systems. Dual-source heat pumps have higher efficiency ratings than air-source units, but are not as efficient as geothermal units. The main advantage of dual-source systems is that they cost much less to install than a single geothermal unit, and work almost as well.

Even though the installation price of a geothermal system can be several times that of an air-source system of the same heating and cooling capacity, the additional costs may be returned in energy savings in 5 to 10 years, depending on the cost of energy and available incentives in your area. System life is estimated at up to 24 years for the inside components and 50+ years for the ground loop. There are approximately 50,000 geothermal heat pumps installed in the United States each year. For more information, visit the International Ground Source Heat Pump Association.

These systems, though small, on the order of 100,000–500,000 btu/hour are directly applicable in principle to those used in larger industrial applications.

Heat pumps can also be used to heat swimming pools as well as provide domestic hot water.

3.24.2 As a Source/Sink for Large Industrial Loads

In HVAC applications, these are both cooling or heating based. This is possible in subsurface media with relatively high thermal conductivity. Care must be taken to avoid thermal overload.

For CPI heat pump applications, technology is employed above ground in integrated power and production environment where optimal heat recovery and power generation can be force-fit designed. High COP, in excess of (5/1) benefits arise from constant process operating conditions similar to the benefit for ground source HVAC heat pumps that tap into a constant subsurface earth temperature. But when CPI heat pumps are used in ground source mode, loads need to be carefully designed and matched to not overburden local thermal profiles and the ability to dissipate or withdraw heat. An example comparing the economics of heat pumps to furnaces is in the next chapter (Chap. 4).

For both HVAC and process applications, ground source well spacing and number are critical as ultimately $Q = U \times A \times$ DeltaTLogMean where the overall heat transfer coefficient (*U*oa) is not very high and must be balanced against temperature and areal driving force.

And remember, the overall CPI approach will be centered on integrated power and chemical production, so some waste materials will become feeds within such facilities. Some waste heat will become preheat for other processes.

Mechanical vapor recompression is employed in distillation, particularly evaporators. But the vapors of one distillation column can be compressed and fed into the reboiler/bottom of another column separate from this process. Also, high volume low-quality waste heat can be evaporated and compressed, under appropriate conditions, to create a high-quality heat source to improve overall process thermal efficiency. Process economics will dictate these applications.

3.25 Other Interesting SE Approaches

Gasifier-based ammonia production [Ch7:4] has been proposed as a method to store energy. Separation into hydrogen and nitrogen as a fuel is seen as a CO_2-reduced methodology.

Pumped water can be used as energy storage methodology through off-peak pumping to elevation followed by on-peak generator let down. The approach is applicable to either a 12 h/day operation or for storing excess energy for peak use. If enough real estate is available, could power 24 h production.

3.26 Summary

Ultimately, the sun is our key to a totally sustainable future. All food ultimately comes from the Krebs cycle, aerobic and anaerobic metabolic pathways powered by the photosynthetic processes. There is no magic here. The amount of material we remove from the earth must be in balance with what goes back in. I like the farm as an example. The sun seems to be glad to supply an infinite, relatively constant amount of energy. If CO_2 and O_2 can be kept in balance, things will be pretty good. Once again, we come full circle to and face to face with the material and energy balance. Fact based always.

References

1. US Energy Information Agency. Tables of US Energy utilization. www.eia.gov
2. Goodwin DK (1994) No ordinary time: Franklin and Eleanor Roosevelt: the home front in World War II. Simon and Schuster, New York
3. Aguayo R (1991) Deming's forward in "Dr. Deming—the man who taught the Japanese about quality". Simon and Schuster, New York
4. Crosby P (1979) Quality is free. McGraw-Hill, New York
5. Exergy Paper from DOE for process stream energy availability—exergy analysis: a powerful tool for identifying process inefficiencies in the U.S. Chemical Industry. Summary report December 2004, Study conducted for the U.S. Department of Energy by JVP International, Incorporated and Psage Research, LLC
6. Stamicarbon. Fertilizer production from gasification. www.stamicarbon.com
7. Midrex. Iron reduction form gasification. www.Midrex.com
8. Worrell E, Galitsky C (2005) DOE/EPA report. Energy efficiency improvement and cost saving opportunities for petroleum refineries An ENERGY STAR® guide for energy and plant managers Ernest Orlando Lawrence Berkeley National Laboratory, February 2005
9. U.S. Energy Information Administration/Annual Energy Review (2011) Table 1.3, P9 and table 10.1, P279. www.eia.gov
10. caltech.edu/news/artificial-leaf-harnesses-sunlight-efficient-fuel-production-47635

Chapter 4
The Efficiency of All Things

4.1 Efficiency in Our World: Theory Meets Practice

As a young engineer, the concept of efficiency took me quite by surprise—which turned into amazement followed by anger that eventually calmed down to understanding and focus on how to improve things. That a seemingly straightforward process design on paper that would cost, say $100 dollars to purchase equipment, would in fact cost as much as $500 to install and get up and running when all was said and done was bad enough. I soon learned that energy to operate pumps, motors, columns, etc., the actual delivered power, would cost anywhere from two to five times more than my theoretical calculations. This quickly brought home the realization that in addition to the oversimplifications of, for instance, the ideal gas law and other theoretical limiting physics concepts, the equipment itself often had other limitations on its ability to convert energy into useful work. The importance of comparing theory versus practice through use of adjustable or, tunable models, e.g., Aspen, ChemCad, and others, in order to better represent a process also became apparent. This also serves as an excellent double-check in either design or operations mode. (Again, the double-entry system checklist).

How, I puzzled, could this be? I started to get an inkling of this while a senior undergraduate engineering student, but it really hit home when I went to work in an operating chemical plant that used fractional distillation, a very energy-intensive process with wasteful once-through cooling and condensing as its primary means of production. I soon also became aware that anything that turned electrical, chemical, or mechanical energy into useful work would not approach 100% efficiency and rarely got above 50 %. Chemical engineering thermodynamics, and I like B.G. Kyle's referral to "Process Thermodynamics" [1], that informs us this lost energy showed up principally as heat, or untapped chemical potential. Of course, science-based thermodynamics, which serves as the underpinning of its process engineering

J. P. Perl, *Sustainability Engineering for Enhanced Process Design and Manufacturing Profitability*, https://doi.org/10.1007/978-3-031-52363-2_4

counterpart, constantly reminds us that entropy is increasing, or, in the limit of no work being actually done, is breakeven at best.

Later with more practice in industrial design and operations, consulting, and teaching of design, it became apparent that the process of turning raw materials into finished products in the chemical industry could incorporate and even make use of some of the heat-generating efficiency limitations through careful design. This is, in fact, an important SE design tool, if not a downright requirement, even more so if power generation could be incorporated.

There are various ways to calculate efficiency and energy utilization including availability, as shown by Callen [2], and exergy (Chap. 3, Ref. 5). A Department of Energy (Chap. 3, Ref. 5) commissioned study presents a review of exergy analysis that presents a methodology for getting at useful extractable process work.

Through sustainability engineering design principles, the chemical process industry can make additional good use of what is referred to as low-quality heat in ways unavailable to Carnot engines. In the power business, the concept of combined heat and power (CHP) generally refers to combining internal steam production with electric power generation, e.g., the HRSG of Chap. 3. In Chap. 8, we look at some examples involving the judicious combination of disparate energy and material conversion processes to reduce cost, increase human and environmental safety, maximize sustainability, and increase profitability. This will involve combining energy production with the manufacture of various different products, chosen to maximize the transfer of energy, be it cooling or heating, not commonly possible in traditional single-product production lines.

Keep in mind that transportation of natural gas to the end user consumes about 10% of its energy. Conversion and transportation of fossil fuels to electricity can consume 67 % just to get electricity to the plug. This allows the judicious use of fossil fuels for onsite electricity generation as well as direct production from chemical or gasified organics in the form of H_2, CO, and CO_2.

4.1.1 Good Old Einstein: $E = mc^2$

I was going to leave this chapter focused on energy alone, but it is worth remembering that the first course of study for budding chemical engineers is material and energy balances. This is how we "check our work," i.e., the conservation of matter and energy, requires namely neither created nor destroyed, but take new forms. Also, SE demands, among other things, the simultaneous consideration and optimization of the use of both materials and energy consumed during production.

Some process design considerations for material conservation include:

1. Better reaction chemistries, e.g., biological, catalytic, molecular, enzymatic
2. Better liquid–liquid extraction to squeeze the value out of everything
3. Better mixing to enhance molecular diffusion and reaction

Post-reaction considerations for material conservation can include pyrolysis, gasification of by-products to CO, H_2, or NH_3 for reformatting to other useful merchant products, or simply fuel for use elsewhere. This can be extremely profitable if disparate chemical processes are combined that can use each other's waste or reformatted chemicals.

Process design considerations for energy conservation focus on energy efficiency; where power generation and export are unfeasible, look to using reformatted chemicals for onsite power or as a feedstock to the disparate coterminous reaction product streams.

Here is a brief historical look at regulatory energy policy development that SE draws upon.

4.1.2 Relevant US Energy Policy Driving/Affecting Commercial, Industrial, and Residential Energy Utilization

Here are many of the pertinent federal energy policy regulations dating back to the 1970s that kind of got things rolling. Things got a big push with the OPEC Oil Embargo of 1973. This was the first big jolt America got regarding its dependence on foreign oil for energy and its use as a political tool. Sure, we had lots of oil and gas, still do, but our dependency allowed us to "bank" some of our resources for the future. This all changed as a result of the Mideast conflict that year and led to legislation to put USA on the road to self-sufficiency through enhanced exploration and development supplanted by the first automobile mpg fuel efficiency standards for US auto manufacturers.

These are synopsized from the energy.gov website and provided here to give some idea of the past history and future leading to SE regarding energy.

4.1.2.1 Energy Policy and Conservation Act of 1975

Appliance and equipment efficiency standards have served as one of the nation's most effective policies for improving energy efficiency. The first standards were enacted at the state level in California in 1974. At the national level, the Energy Policy and Conservation Act (EPCA) was enacted in 1975, and established a federal program consisting of test procedures, labeling, and energy targets for consumer products. EPCA was amended in 1979 and directed the Department of Energy (DOE) to establish energy conservation standards for consumer products.

4.1.2.2 The Naval Petroleum Reserves Production Act of 1976

From energy.gov:

For much of the 20th century, the Naval Petroleum and Oil Shale Reserves served as a contingency source of fuel for the Nation's military. All that changed in 1998 when Naval Petroleum Reserve No. 1, known as Elk Hills, was privatized, the first of a series of major organizational changes that leave only one of the original six Federal properties in the program.

Set aside in a series of Executive Orders in the early 1900s, the government-owned petroleum and oil shale properties were originally envisioned as a way to provide a reserve supply of crude oil to fuel U.S. naval vessels in times of short supply or emergencies. The Reserves remained mostly undeveloped until the 1970s, when the Nation began looking for ways to maximize its domestic oil supplies. In 1976, Congress passed the Naval Petroleum Reserves Production Act authorizing full commercial development of the Reserves. The crude oil, natural gas, and liquid products produced from the Reserves were sold by DOE at market rates. Revenues were deposited to the U.S. Treasury.

One of the largest of the Federal properties, the Elk Hills field in California, opened for production in 1976 and became the largest (in terms of production) oil and natural gas field in the lower 48 states at one point in its history. In September 1992, the field produced its one billionth barrel of oil, becoming only the thirteenth field in the Nation's history to reach that milestone. While managed by DOE, Elk Hills generated over $17 billion in profits for the U.S. Treasury.

See energy.gov for additional status.

4.1.2.3 The National Energy Act of 1978

Excerpt from the National Energy Act of 1978:

(a) ***Findings***

The Congress finds that—

(1) *the United States has survived a period of energy shortage and has made significant progress toward improving energy efficiency in all sectors of the economy;*
(2) *effective measures must continue to be taken by the Federal Government and other users and suppliers of energy to control the rate of growth of demand for energy and the efficiency of its use;*
(3) *the continuation of this effort will permit the United States to become increasingly independent of the world oil market, less vulnerable to interruption of foreign oil supplies, and more able to provide energy to meet future needs; and*
(4) *all sectors of the economy of the United States should continue to reduce significantly the demand for nonrenewable energy resources such as oil and*

natural gas by implementing and maintaining effective conservation measures for the efficient use of these and other energy sources.

Excerpt from the Office of Scientific & Technical Information Technical Reports:

The National Energy Act of 1978 contains many provisions that will significantly affect solar technology commercialization and solar energy users. Four of the five statutes that comprise the National Energy Act deserve close attention. The National Energy Conservation Policy Act will promote residential solar installations. The Energy Tax Act will accelerate both residential and commercial solar system applications. The Public Utilities Regulatory Policies Act promotes efficient use of utility resources as well as decentralized power production. And, the Power Plan and Industrial Fuel Use Act places severe restrictions on future burning of petroleum and natural gas, which should lead some operators to build or convert to solar energy systems. Each of the preceding acts are considered in separate sections of this report. Federal regulations issued pursuant to the various provisions are also identified and discussed, and some of the problems with the provisions and regulations are noted.

(Also from https://www.energy.gov/sites/prod/files/2017/09/f36/DOE%20 1977-1994%20A%20Summary%20History_0.pdf.)

A very interesting timeline is presented including:

November 9, 1978 President Carter signs the National Energy Act, which includes the National Energy Conservation Policy Act, the Powerplant and Industrial Fuel Use Act, the Public Utilities Regulatory Policy Act, the Energy Tax Act, and the Natural Gas Policy Act.

4.1.2.4 National Appliance Energy Conservation Act of 1987

Excerpt from energy.gov:

The National Appliance Energy Conservation Act of 1987 established minimum efficiency standards for many common household appliances. Congress set initial federal energy efficiency standards and established schedules for DOE to review and update these standards. The Energy Policy Act of 1992 (EPAct) added standards for some fluorescent and incandescent reflector lamps, plumbing products, electric motors, commercial water heaters, and heating, ventilation, and air conditioning (HVAC) systems. EPAct also allowed for the future development of standards for many other products.

For the complete text, see https://uscode.house.gov/view.xhtml?path=/prelim@title42/chapter91&edition=prelim

4.1.2.5 Energy Policy Act of 2005

Excerpt from energy.gov:

In 2005, the Energy Policy Act (EPAct 2005) set new standards for 16 products and directed DOE to set standards via rulemaking for another five.

Excerpt from epa.gov:

The Energy Policy Act (EPA) addresses energy production in the United States, including: (1) energy efficiency; (2) renewable energy; (3) oil and gas; (4) coal; (5) Tribal energy; (6) nuclear matters and security; (7) vehicles and motor fuels, including ethanol; (8) hydrogen; (9) electricity; (10) energy tax incentives; (11) hydropower and geothermal energy; and (12) climate change technology. For example, the Act provides loan guarantees for entities that develop or use innovative technologies that avoid the by-production of greenhouse gases. Another provision of the Act increases the amount of biofuel that must be mixed with gasoline sold in the United States.

4.1.2.6 Energy Independence and Security Act of 2007

Excerpt from energy.gov:

In 2007, Congress passed the Energy Independence and Security Act (EISA 2007), enacting new or updated standards for 13 products. EISA also included a requirement that DOE maintain a schedule to regularly review and update all standards and test procedures.

4.1.2.7 Energy Improvement and Extension Act of 2008

Excerpt from the Alternative Fuels Data Center, *enacted on October 3, 2008:*

"The Energy Improvement and Extension Act of 2008 is Division B of the Emergency Economic Stabilization Act (Public Law 110-343(PDF)). Title II of Division B of the law includes several provisions related to tax credits and exemptions for alternative fuels and fuel-efficient technologies. The table below provides a summary of the relevant provisions." Authors note: The table covers biodiesel, energy tax credits for plug-in electric vehicles, and various other extensions e.g., electricity as an alternative fuel.

4.1.2.8 The American Recovery and Reinvestment Act of 2009

Excerpt from the Alternative Fuels Data Center, enacted on February 17, 2009:

Provides $3.2 billion toward the Energy Efficiency and Conservation Block Grant Program, as authorized by the Energy Independence and Security Act of 2007. These grants are for state, local, and tribal governments for energy efficiency improvements, primarily in the transportation and building sectors.

Provides $2 billion toward grants for advanced battery systems and electric vehicle components manufacturing. These funds are intended to support domestic manufacturing of advanced lithium ion batteries and hybrid electric systems and components. Provides $300 million toward competitive grants for alternative fuels and advanced vehicle projects, as authorized by Section 721 of the Energy Policy

Act (EPAct) of 2005. The grants are for state governments, local governments, and metropolitan transportation authorities, in partnership with an active and designated Clean Cities coalition.

Provides $400 million to support vehicle electrification efforts. Refer to Idaho National Laboratory's Light-Duty Electric-Drive Vehicle and Charging Infrastructure Testing - Plug-in Electric Vehicle and Infrastructure Analysis Report(PDF) for more information.

4.1.2.9 Consolidated Appropriations Act of 2021: Modernizing US Energy Policy

Excerpt from energy.senate.gov, the Energy Act of 2020:

This legislation includes programs to develop and deploy renewable energy, improve the efficiency of our homes and businesses, modernize the grid, reduce carbon pollution from industrial and traditional power sources, and more.

4.1.2.10 The Inflation Reduction Act of 2022 (Largest Energy Investment in US History)

Excerpt from energy.gov:

Energy portion focus is on the development of domestic clean energy manufacturing and loans for the Energy Infrastructure Reinvestment Program.

4.2 Example Efficiency Standards Mandate

The efficiency of all things has an enormous impact on design of all types. SE designers should always be informed and guided by this knowledge and not accept business as usual, nor take the lowest initial cost while losing a lifetime of energy savings. SE efficiency standards for process design engineering might be needed to protect project owners and engineers.

Two good SE examples with huge commercial impact are the automobile and home utilities. I present these because they are ubiquitous and volumes of data are available for exemplary engineering evaluation. This makes them good quality models to emulate in industry as well, i.e., always demand fact-based technical information and then provide metrics to evaluate actual, measured, versus predicted effects before going forward. Tables 4.1 and 4.2 are representative of mandated efficiency improvement in the private sector for cars and furnaces, respectively.

→

Table 4.1 The fleet average MPG mandates for the US auto industry (approximate values)

1970	10 mpg baseline
1985	25 mpg set in the mid-1970s
2025	54 mpg set recently

Table 4.2 Energy star requirements for home heating furnaces (notional numbers)

Old gas furnace efficiency	60–70%
New minimum requirements	78% (80% by Nov 2015)
Optional: Condensing Furnace	90–95%

A few years back, EPA revised their mpg ratings to reflect 65 versus 55 mph highway speed limits, as well as more realistic test conditions.

✈

When these were first proposed in the mid-1970s, there was much gnashing of teeth, but because of this we all enjoy enormous benefits in reduced reliance on nonrenewable resources, e.g., petroleum and in particular, foreign sources energy (depoliticizing effect). Coupled with the EPA Clean Air Act (CAA) regulations that gave us catalytic converters and lower unburnt hydrocarbons and oxides of nitrogen emissions, we now possess a much cleaner and cheaper to operate automotive environment. These mandates also led to more dependence on electronic ignition, fuel injection and multi-speed and continuously variable transmissions and hence development of more efficient as well as reliable systems. These electronic fuel injection and ignition systems require tighter combustion process control, but produced enormous human health and economic benefit. Improved reliability, almost as a direct consequence of these modifications, helped increase auto sales as well so everyone wins. Why does this happen? Again, every time engineers get a chance to revisit a design, they generally find ways to improve everything about it, not just efficiency.

Consumers around the world are now reading the product stickers, demanding higher fuel economy, and willing to pay for it. This also led to hybrid and electric car development and probably a bridge into research-developed propulsion methods that are not yet devised.

So how do the MPG improvements compare with the increased costs, if any? The quick answer is this was a terrific win for the consumer as well as auto producers who now had yet another marketing tool to increase sales to those wishing to better their MPG and willing to shed their old cars for newer more efficient ones. (Law of unintended consequences strikes again, in a positive manner)! Over the past 30 years or so, the average new car warranty jumped from 12 months to 36 months and in some cases well beyond that, economically supportable largely with a more reliable product. Look for longer guarantees as well when specifying and selecting process equipment.

✈ Always verify the energy efficacy of gasoline versus hybrid versus pure electric, as it does not always make sense, and the same logic applies to process resource utilization evaluation. ✈

The presence of Energy Star product labeling for oil- and gas-fired devices provides the consumer with plain, easy-to-understand information on the value of investing beyond the 80% combustion efficiency range. The additional upfront investment costs are typically recovered in 5 years or less compared to the 20–25 year life of properly installed equipment. In Chap. 7, we look at the use of industrial process heat pumps to achieve on the order of 10× multiplication of energy utilization in integrated power and chemical production manufacturing facilities. And remember, theory tempered by efficiency moderated by actual experience = improved performance!

The vision to require these across the board efficiency improvements was initially motivated by the 1973 OPEC oil embargo and a desire to decouple the US from politically imposed controls on supply of this critical resource once taken for granted that would respond to classic economic supply and demand. This government mandate has been successful beyond any of the wildest predictions and arguably moved us into the green era of consumerism. SE design will do the same and for the same reasons. Another positive unintended consequence is the enormous reduction in CO_2 footprint resulting from a doubling of automobile fuel efficiency. And do not forget CO_2 released to the atmosphere is lost carbon which could have been used for more valuable production, e.g., plastics, so keep carbon where it is useful.

✈ Electric resistance heating is limited, i.e., defined by Ohm's Law, but the use of air source heat pumps can reduce electricity use by 50–75% for HVAC and more when ground source (GS) geothermal is employed. No magic here, just a bit of thermodynamic jujitsu that leverages/elevates low-quality heat. The concept is analogous to a catalyst that lowers an activation energy barrier to promote a chemical reaction. The same total energy conservation principle is employed in a CPI plant environment. ✈

✈ We are lucky the moon is not inhabitable or unsustainable capitalism would simply deplete the earth and its resources and then move on to the moon! Note also that SE done properly is the new enlightened capitalist's tool and will save money all around therefore meeting stockholder as well as consumer expectations. Just ask Whole Foods and their customer base. Just ask Wal-Mart why they care so much about the energy footprint of the products they sell (see Chap. 3. Sect. 3.2). (Helium for fusion reactor just announced so maybe we need the moon after all, where it recently was announced to be plentiful!) Enlightened self-interest … good for the consumer … good for business … good for the planet. ✈

There are many ways to make theoretical calculations regarding energy utilization. The most common method of evaluation and comparison involves some form of empirical determination of conversion efficiency of available mass and energy. All the theory is useless unless compared with experience, i.e., data, whereby improved alternatives can be devised if possible. For example, the automobile internal combustion engine (ICE) performs well only over a narrow range of RPM and speed, and even this is pretty poor, ca 10–15% efficiency! The electric motor, on the other hand, has very good efficiency over a much broader range and so is used to boost overall gas mileage by 80–90% particularly during acceleration. As used in hybrid automobiles, this allows the ICE to be more closely tuned to maximize highway mpg.

Since the late 1930s, the railroad locomotive, basically a diesel engine-charged, electric motor-propelled design, has taken even greater advantage of these engine performance characteristics. But even these novel applications still waste tremendous amounts of the stored energy of organic fuels. This wasted energy, in the form of heat, can be used in stationary, integrated power and chemical processes, or local hot water needs by capturing low-quality heat as we shall see.

The following list is presented to serve as a reminder not to assume anything about energy efficiency. Always demand and then test equipment performance guarantees. The key here is the use of energy, not cost alone in evaluating efficiency. It is tempting to choose the least cost upfront, but an NPV should show the way here if properly evaluated. Even with attractive net present (NPV) value, always err on the side of greater efficiency. Also err on the side of higher and flatter efficiency curves that will allow your process greater turndown ratio flexibility and ability to operate with disparate processes. From a controls perspective, always avoid positive feedback coupling that can lead to growing, instead of damping oscillations around the set points for either process(s).

4.3 Some of the More Interesting Fun Facts of Efficiency (Nominal Values)

1. Conversion of food to useful muscle energy to work by humans—20%
2. Horses are somewhat more efficient, with a bio footprint 2.5 times smaller than a tractor
3. Conversion of organic fuels, e.g., coal or natural gas to electricity—delivered to plug, 33%
4. Seven percent of incoming crude oil (BOE) is consumed as energy resource in refinery operation
5. Automotive conversion of gasoline to conventional propulsion: 10–15%
6. Conversion of gasoline to hybrid/gasoline/electric motor automotive propulsion: 20–30%
7. Conversion of oil to gasoline at refinery: 85–90%
8. Conversion of wall electricity to electric automotive propulsion: 40–50%
9. Pipeline delivery of natural gas to point of use 90%
10. Home natural gas furnace, standard, mandated: 78%
11. Home natural gas furnace, condensing: 92–94%
12. GS heat pump central United States, produces 5 Btu/Btu purchased electricity: COP = 5/1
13. Typical fluid pump 60–80%
14. Typical natural gas compressor 70–85%

4.4 Example: Economic Comparison of Ground Source Heat Pump and High-Efficiency Condensing Furnace

Figure 4.1 shows the comparison between a ground source heat pump and a modern, high-efficiency condensing furnace (Basis = 100,000 Btu/h).

When using electricity, one always must factor in the overall efficiency of 33% from oil to plug. Alternately, transport of natural gas is 90% efficient. So far so good, but hold on, NG fracking in the Bakken field, e.g., has reduced the price of NG from 10 → 3$/MMBtu. Confused? Good! And this is an easy one! The heat pump only needs 15,000 Btu of electricity, or [15,000 Btu/h] × [1 kW-h/3413 Btu] = 4.39 kw. At about $0.11/kW-h, that is about $0.48/h to operate.

Our gas furnace costs [(100,000 Btu/h)/0.94] × $5.5/10^6 Btu = $0.59/h, or about 22% higher. The amount $5.5 includes taxes, same as the electric rate, a close call. The heat pump can win in that it also serves as A/C in the summer, and GS typically provides an additional 10–15% cooling application benefit.

A COP = 4, however, can swing economics the other way, same issue with furnace efficiency, so performance guarantees really come into play, as well as cost of a crystal ball to forecast electric and gas rates!

✈ Review and verify the Performance Guarantees here!

✈ Design/Build/Commission/Startup—Check at each step of the way!

Process heat pumps must always be held to the same level of scrutiny. Figure 4.1 shows GS heat pump compared with modern high-efficiency natural gas fired condensing furnace. And remember, in a process setting, the condensing section might be another process that needs heat at lower temperature, hence the importance of siting disparate processes. If safety is a concern, a buffering heat transfer fluid, e.g., hot oil, can be placed between combustion area and the process.

4.5 Other Efficiency Review Examples

There is no end to the study of equipment efficiency, but there certainly is a lot of data available. Vendors of pumps, compressors, motors, column packing to name a few, generally provide performance curves from which the designer can optimize efficiency. A few excellent textual references include:

1. Working guide to process equipment
2. Distillation design
3. Distillation control

USEPA and DOE as well as ASHRAE rate various sorts of mechanical equipment.

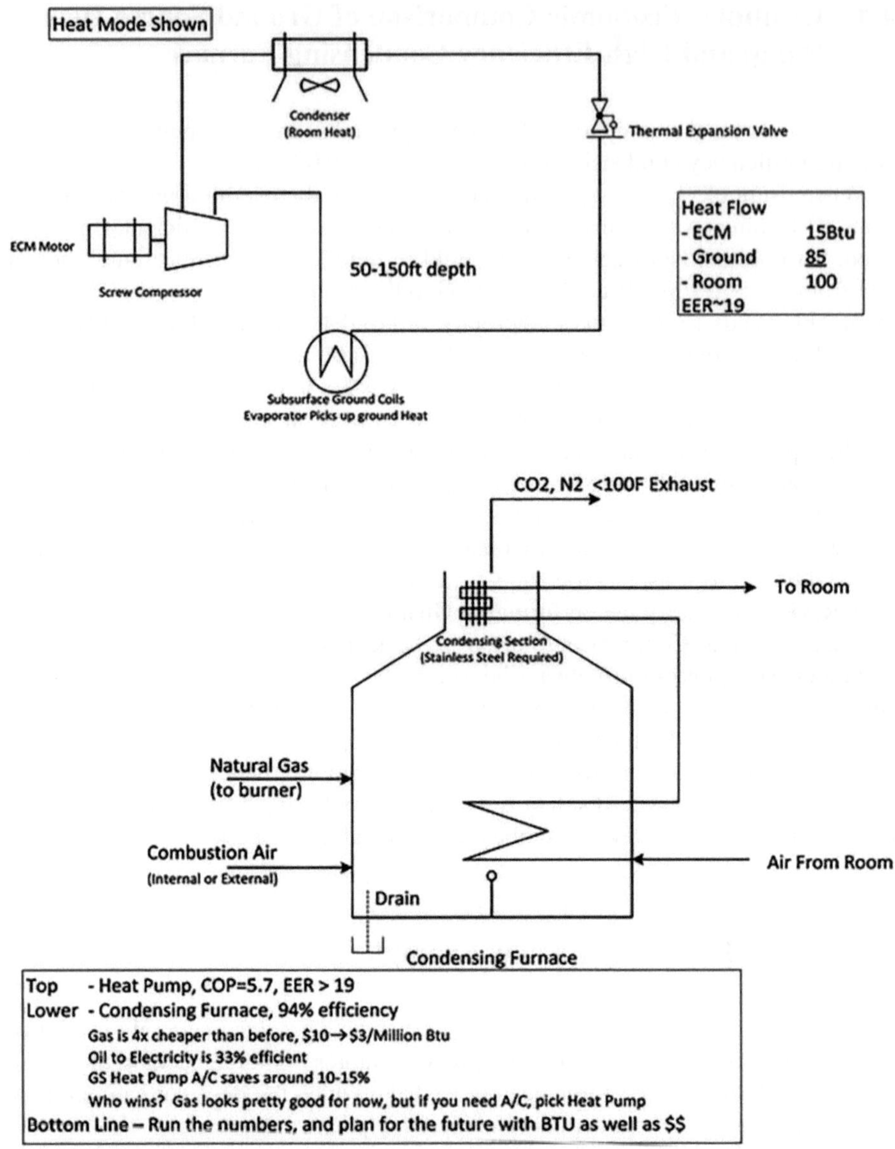

Fig. 4.1 Ground source heat pump HVAC application comparison with high-efficiency condensing furnace

4.6 Gas/Hybrid/Full Electric Vehicles

Even though a car is a far cry from a piece of process turbo-machinery, it is still worth looking at this simple example because so much data is available, and this is after all a method of converting fuel into motive power. The Alternative Fuels Data Center (AFDC) [3] reports that 1 gallon of gasoline (10% ethanol content) has a lower heating value (LHV) of approximately 114,000 Btu/gallon and a higher heating value (HHV) of 122,000 Btu/gallon. EPA assumes the energy content for 1 gallon of gasoline to be 33.7 kW-h, or 115,000 Btu, or approximately the LHV of the energy content. This is a reasonable way to evaluate as a car has no ability to do anything with the heat lost in tailpipe water vapor. This is not true of the stationary HRSG and certainly not true when used in an integrated power and chemical plant. These facts also highlight losses associated with a 2000 F+ internal combustion process! (Exergy or availability are probably better ways to evaluate.)

Table 4.3 is not a complete list of vehicles but is representative of what is presently available. It also shows the relative efficiency of each option. Something else to keep in mind in the case of all electric vehicles is electric motor and battery life. At present, there is no realistic upper end life for modern gas engines. They seem to last forever if maintained (fuel, lubricant, and coolant systems). Process designers must recognize this total cost of ownership including maintenance and periodic replacement need when performing economic evaluations of design options.

4.7 Home Furnace and Process Industrial Steam Boilers

Note that a similar efficiency type calculation comes into play with the condensing versus noncondensing home heating furnaces. An 80%-efficiency furnace will have high temperature exhaust vapors that require metal pipes and brick or other insulated chimneys. Modern condensing furnaces, however, capture the HHV in the water vapor while reducing excessive exhaust temperature. This removes the need for metal and chimney exhausts, but leads to four unintended consequences for the designer to consider. These design considerations are typical of any SE design changes for existing process systems and will need to account for:

1. Reduced exit temperature re: loss of natural exhaust draft for chimney operation

Table 4.3 Equivalent fuel efficiencies MPGe (e = equivalent) for some electric cars [4]

Car	MPG by EPA	Notes
Nissan Leaf	114e	Electric
Prius	50	Gas/electric hybrid
Prius Plugin	95e	Gas/electric plug-in
Chevrolet Bolt	120e	Electric
Tesla Model 3	131e	Electric

2. Second-stage condensing heat exchanger is exposed to carbolic acid, a by-product of CO_2 combustion, and liquid water, requiring stainless steel construction
3. Removal of condensed water (similar for air conditioning).
4. Reduced cold weather room air exchange when outside air is used for combustion (two-pipe system), a common method for increasing efficiency to 95+ %.

For the first issue, the low natural draft condition is taken care of by using plastic exhaust piping direct to outside. For the second, acid corrosion is prevented by employing more expensive stainless steel in the secondary heat exchanger section. The third is addressed by a small sump pump to remove the condensed water (also used for air conditioning condensate). Finally, older furnaces draw combustion air from inside the living space creating a slight negative pressure that brings outside fresh air into the home, so consideration of interior space ventilation with high-efficiency two-pipe systems is important as windows and doors may not be open as often in cold weather.

As gas is combusted, it draws air into the home as exhaust gas passes out the chimney. For in home condensing furnace installations, the use of a separate inlet air pipe can increase efficiency from 90 to 94+ . Again, the engineer must consider all consequences and in this case, air coming in from outdoors must be heated to 70°F, e.g., double pipe exchanger. This is lost energy, but it also ensures a flow of clean air into the home during winter months when windows are closed. In a plant situation, this may be of little concern where the name of the game is saving money. The point is all system-specific considerations must be made during the design phase. Install the most efficient equipment upfront or lose 20 years of energy savings.

4.8 Ground Source (GS) Geothermal Heat Pump

The Earth's temperature at between 50 and 150 ft often settles in at about 55°F. Designers of HVAC systems for home or industrial heating and cooling applications can make good thermodynamic use of this fact. In A/C mode, removal of heat from living space to outside is more efficient if the receiving temperature is low and thermal mass high. By placing the heat transfer coils into the ground, these conditions are met, greatly increasing system capacity and efficiency. In the winter, the heat pump gets "free heat" from the earth.

There is an HVAC example in Fig. 4.1, but it should be pointed out that buried heat transfer coils can also be employed in industrial settings so long as the heating and cooling loads are balanced so as not to overload the ground source heat transfer coefficient and capacity. More commonly, in an industrial setting, the role of the buried coil is played by an adjacent thermal sink or source, leading to high COP/EER, as with process mechanical vapor recompression (MVR) systems. Swimming pools can be heated efficiently but not quickly in the same manner. See Chap. 7, Sect. 7.13 for a heat pump in process analog.

4.9 Onsite Power Production in CPI Facilities: An SE Efficiency Booster

Onsite energy production can save at least the 25–30% electric transmission line loss (10 % for natural gas) between power station and end user. In addition, reuse of waste heat from HRSG is also applied to savings in combined power and chemical production facilities. Again, this is due to the CPI facility's ability to use both high- and low-quality heat. A downside of course is the need to maintain power generating equipment and to balance process needs via grid connection. This is not a reason not to do it, but rather a reason to learn how to manage it as part of the process. Again, the power generation final arbiter is overall economics.

Hidden sustainability issues impacting efficiencies not often stated in Btu and materials footprint analysis include transportation, mining, and government subsidies to name a few. These are important as most energy prices are quite volatile, while end use process Btu requirements remain relatively constant. For the purpose of long-term sustainability planning, the production of a material or form of energy should be stated as the amount of energy it requires per unit sold, similar to a nameplate heat exchanger duty statement. Remember, the efficiency of all equipment must be maximized upfront to assure long-term SE performance. Overly simple cost evaluation may not meet SE requirements.

Chemical process production utilization of feed materials also has a sort of associated efficiency, similar to energy. In pharmaceuticals, for example, tree bark may be extracted to obtain the salicylic acid used to create the active ingredient found in aspirin, but with an enormous generation of waste by-products. Modern, scientifically, acceptable methods of harvesting trees for such use or even for use as lumber must be done sustainably if for no other reason than to assure permanent access to such valuable materials. Materials cost is a short-term, volatile factor. Renewability assures not only availability, but can serve as a good business hedge against cost increases, i.e., less supply and demand effect.

✈ There are other synthetic methods to make aspirin, but the original one provides a good example of material utilization regarding SE. ✈

Chemical reactor selectivity and conversion also provides an example of a process material production efficiency. Low efficiency here can be addressed by co-locating disparate processes that can utilize unconverted or reactor products as feedstocks.

By EPA regulation, chemical production by-products are not wastes if they can be used as feed for another process. This suggests an additional reason/benefit to collocating unrelated, but complimentary production processes in an integrated fashion.

4.10 Common Hierarchy of By-Product Utilization[5]

This would include:

1. Use in a follow-on process, e.g., paper production
2. Bio-fermentation of garbage and municipal solid waste to generate methane for either chemical production or transportable fuel
3. Gasification to make heat and/or power to support production
4. Seek out material and waste exchanges first
5. Landfill as a last resort if allowed

When examining alternatives, industrial chemists need to weigh pharmaceutical efficacy against cost and sustainability of supply. For example, synthetic methods of aspirin production have replaced the multi-thousand year old method of tree bark utilization for pain relief. Over the past 20 years, EPA developed the Ecological Structure Activity Relationships (ECOSAR) model (Chap. 5), a computerized chemical evaluation program to assist in the design of low to no toxicity alternatives for desired new chemical product use. The gist of this revolves around making minor changes such as number of double bonds in a molecule that has minimal effect on desired functional properties while greatly reducing or eliminating toxicity. Of course, when contemplating process changes that involve FDA approval, great care must be taken as manufacturing permits are typically based not only on the materials but also the preparation and manufacturing procedures used. Drug efficacy can be highly dependent upon both factors.

All design engineers are familiar with the concept of efficiency, but not all have had the opportunity to venture beyond the traditional. Most of what we do costs a great deal of money to implement so as a breed we tend to be conservative. Utilizing low temperature, low-quality waste heat to improve thermal efficiency is quite common, but integrating power generation and production of unrelated chemicals is not; so, we will take a look at this sustainability-boosting method now.

4.11 Combined Heat and Power (CHP): Efficiency in the Chemical Process Industry

CHP is practiced within the power industry and the HRSG described in Chap. 3 is a good example of this. In this book, we start referring rather to "integrated power" to differentiate existing CHP from a newer SE approach that integrates power production with chemical processing.

4.12 Electric Power Generation

Conversion of fossil fuels into electricity is not an efficient process. Typically, only 30% of electricity is delivered to the plug. But, by using heat recovery, in HRSG, this can be improved to ca 40%. Thomas Edison used waste heat in a district heating scheme in one his earliest New York City area power stations ca 1890 [6]. But an even greater improvement in efficiency may be had by combining electricity generation and chemical production in an integrated chemical process environment. The goal here is to maximize energy conversion and minimize material consumption thereby increasing sustainability.

4.13 Power Generation Integrated with Chemical Production: A Key SE Factor

In the past, the best chemical manufacturing facilities always found ways to use as much of the wasted process energy in heat recovery exchangers, known as economizers. This practice is so widely used, e.g., in refining, that a new heat exchange network (HEN) evaluation method known as pinch technology [7] was developed to make sure the energy gradients, normally hot to cold, did not "cross." Here, a cold feed stream flows against a warm product or intermediate stream. In this manner, significant energy can be recovered from the warm stream, but not all. By generating electricity onsite via IGCC HRSG, the plant can use virtually all remaining lost Carnot heat for chemical reactions. Using this low-quality heat often requires co-locating disparate manufacturing streams, e.g., ammonia, natural gas liquids, and iron ore reduction to name a few. One example uses natural gas to generate electricity as well as gasifier feedstock to provide the raw materials, CO, and H_2, for further processing into numerous merchant products, e.g., ammonia-based fertilizers, ethylene, pipeline quality gas, and local electricity (see Fig. 8.6).

4.14 HVAC as a Model for Rating SE Efficiency Improvement

After the energy crisis of the mid 1970s, in addition to automobile fuel efficiency standards and speed limits, numerous other efficiency standards were put in place particularly surrounding home natural gas and electricity consumption by all manner of appliances.

AFUE stands for annual fuel utilization efficiency. Space heating through combustion, e.g., natural gas, has been a fairly inefficient energy consumer at 50–70%

AFUE (Btu recovery). Federal laws now require a minimum of 80%, with numerous energy tax incentives to encourage further improvements. This brought about the modern condensing furnace, capable of attaining 94+% efficiency based on the higher heating value of the fuel. But this is possible only by the addition of secondary heat recovery equipment to cool combustion gases to below 212°F. The unintended consequence facing the design engineer is the presence of CO_2 in liquid water that produces carbonic acid, requiring the use of more expensive stainless steel construction. Here is a perfect example of balancing cost against recovered Btus and in the case of mass produced equipment, the benefit is there, but tax incentives are required to offset higher homeowner initial investment cost. Over the past 30+ years, heat pumps have been used that are basically compression-based systems reversible to provide cooling in the summer (100°F) and heating in the winter in mild, >20°F, climates. The need to operate at wide temperature ranges reduces efficiency, much like the internal combustion engine that works much better at constant RPM and load.

A more recent addition to the HVAC toolbox is the geothermal-based heat pump. Here, heat transfer coils are buried in the ground from 25 to 100 ft where the earth's temperature remains relatively constant, at say 55°F. A thermal heat transfer liquid, e.g., methanol or propylene glycol is recirculated to the buried earth coils to alternately provide a heat source and or cooling sink as needed. Unlike air source heat pumps, ground source systems can be run in extremely cold as well as hot climates, greatly increasing efficiency over air source ones, so long as reservoir heat load capacity limits are maintained.

4.14.1 HVAC Standards

In all HVAC applications, the development of government-required SEER (seasonal EER) and EER standards have allowed consumers a broader, understandable method of balancing operating energy savings against higher initial costs. Basic air conditioning systems must now meet or exceed SEER 13 levels, and heating performance for heat pumps require baseline COP of ca 4.0. In addition, tax incentives exist as well to offset initial equipment cost differential. Local utilities have also added investment incentives for replacement of old equipment with more efficient ones. This is very wise as it is cheaper for utilities to reduce electrical demand than to build new power plants.

4.14.2 Compressor Technology Efficiency Improvement

In order to meet the EER the HVAC industry introduced the thermal expansion valve system to replace the highly effective, but limited efficiency expansion orifice. In addition, the two-stage compressor was developed that improves efficiency in much the same way that an automobile multi-speed transmission does. A more recent addition is the variable speed compressor, analogous to the CVT transmission employed in electric hybrids such as the Prius, providing a nearly perfect, more constant match of power and heating/cooling load.

4.14.3 Blower and Pump Motor Efficiency Improvements

The space heating and cooling transfer occurs in devices known as air handlers. Here, living space air is passed over a coil which either cools or heats it. The common motor for this process, permanent split capacitor (PSC), is very useful and reliable, but efficient only over a narrow range of air pressure drop. The more recent advent of electronically commutated motors (ECM) has greatly increased the efficiency by again providing a more continuously variable air moving method. Ground source geothermal heat pumps have started to incorporate variable speed heat transfer loop flow to more perfectly match heating/cooling source and load. All of these advances increase efficiency by providing more constant load matching between sink and source.

4.15 Process Equipment Efficiency and Performance Curves: Read This Before You Purchase!

All design and operating engineers either know this or learn the hard way. Carnot notwithstanding, it is imperative to design your system with complete knowledge of all possible operating ranges to be reasonably expected. When specifying pumps and impellers in particular great care must be taken, especially when large numbers of them will be used, as in refineries. Once installed, pumps and other mechanical equipment's lost energy will be permanent and often difficult or impossible to recover. The lost energy shows up as heat; so, take care to add bypass coolers for large pumps to prevent seal loss if accidently valved-off while running. Better yet, choose VFD controls to prevent that from even happening. And remember, the heat given off in our HVAC is a good thing in the winter, but bad in summer, so it is best just to remain efficient under all conditions.

4.15.1 Efficiency Improvements as an SE Energy Extender

As an example, power utilities around the country have been assisting both residential and commercial customers in finding ways to use less of their product. Early this year my utility came out and did energy studies and exchanged old incandescent and even fluorescent light bulbs with modern LEDs. About 10 years ago I was contracted to the same utility to assist in energy audits of their commercial clients' energy use and to identify ways to reduce electric consumption.

The why of this was easy: cost avoidance of building new, expensive power generation facilities. This was also augmented by federal energy conservation funds.

4.15.2 Efficiency Improvements as an SE Material Extender

Finding new manufacturing chemistries or ways to process the same materials into finished products with reduced feedstock is also an SE extender, here for materials. Yes, companies are incented to do this on their own, but often lack the R&D funds to jumpstart this process. State and Federal funded Government University programs can provide.

There is no better way to make improvements than during new plant construction, or refinery turnarounds. It is always cheaper to install at that time than to shut down periodically, randomly.

4.16 Economics of Process Efficiency

The government has provided product mandates for automobile fuel efficiency, HVAC, refrigerators, etc. This is more than mere meddling, as this equipment all has lives of 10–20+ years. Without these mandates, required across the board, there would be little to no incentive for consumers to pay extra upfront, even though they save in the long run. Also, these improvements, once placed into mass production, become considerably less expensive and more reliable as engineering evolves.

In the CPI, however, it is up to the designer to pick the most SE-appropriate equipment and this will require careful engineering economic analysis coupled with a demanding specification to encourage vendors to produce such equipment that will reduce Btu footprint permanently and evenly across the desired range of operations. Now this has always been the purview of the company to decide how to optimize, but there may be room for Energy Star type standards/ratings from which to select for process equipment to improve the availability and selection. Also from my teaching, I have found the concept of efficiency, beyond Carnot, is not necessarily covered well in academia. Clients who want to spend as little money upfront with an eye toward a 2- to 3-year breakeven points may need some additional incentive/requirement, again across the board, to help SE lead the way. Ultimately, this saves much money as well as resources in the long run. SE intensity can lead the way by demanding the technology for the future.

4.17 Key Item Needed: An SE Equipment Efficiency Rating, a Sort of SE Energy Star Rating

✈ The basic information is commonly available as most equipment vendors provide operating curves. Hint: If they do not, stay away! ✈

4.18 Distillation: The Classic Energy Sink and Source

Distillation is perhaps one of the most energy-consuming of all CPI technologies. The industry needs little encouragement to improve this and has worked on this for well over the past century. Column packing efficiency improvements of a few percent can lead to large savings in refinery settings and this is a good place to look for methodologies of efficiency improvement.

Distillation works by counter current contacting of boiling, rising warm vapors against falling cool liquid. This is the essence of the technology and heat provides the energy driving force while concentration gradient provides the purification and separation driving force. Two primary areas of improvement have seen great advances (See the references of Kister [12], Lieberman [13] and Shinskey [14]).

4.18.1 Contacting Trays and Internals

As a young engineer in training, one of my first lab assignments was to determine the tray efficiency of a new packing. Two principal methods of contacting here include discrete trays and packing that represent equivalent trays. Bubble cap, valve, and perforated trays are prime examples of discrete trays. Two types of packing are structured and dumped. Primary characteristics for packing are pressure drop, height equivalent to a theoretical plate (HETP), number of transfer units (NTU), and throughput of both liquid and vapors. For the case of trays, HETP is replaced by number of theoretical plates as modified by tray efficiency.

The amount of money spent on new contacting method development is enormous as more efficient trays and packing reduce energy requirements and maximize production. Particularly, in the area of petroleum refining, economical throughput increase is the holy grail of refinery managers and has great sustainability effect on reducing carbon footprint as well.

4.18.2 *Energy Reduction Approaches in Distillation Efficiency Improvement*

A relatively new methodology of distillation energy recovery is the divided wall column (Chap. 7, O'Brien et al.). This creates an internal heat exchanger within the column to greatly enhance internal heat transfer. This requires either incorporation into new design and construction, or modification during plant shutdowns or turnarounds. Up to 30% energy reduction may be had. We take a closer look at this in Chap. 7.

All distillation requires heat input for vapor production and heat removal for condensation to liquid. By judicious placement of distillation columns operating at different temperatures and pressures, one tower overhead can be made to condense against another tower's bottoms. This removes the necessity for one condenser and one reboiler. When this is not possible, and if temperature differences from top to bottom are not too great, an overhead compressor can be used to increase the heat and therefore thermal potential energy availability for use as a reboiler. This is referred to as mechanical vapor recompression (MVR) and is basically, a heat pump.

Heat pumps for distillation and evaporation—Basically this is a universal method for changing the thermal potential energy level of a vapor stream to make it available for reuse. Energy savings of greater than 30–80% can be had at the expense of additional equipment. This is described in further detail at the end of Chap. 7.

4.19 Fans Are Not Air Conditioners

Fans and the motors that run them are all subject to efficiency limitations. When I was a young boy we did not have air conditioning and my parents placed a fan in our apartment window. This reversible fan would exhaust hot air during the day and bring in cool air at night to at least provide a little sleeping relief. Thermodynamics and efficiency limits though produce heat, so using a fan inside on the floor next to you could have unintended consequences as it did in Chicago about 5 years ago during a heat wave that killed hundreds. Many of these folks, mostly elderly and poor, were found with windows closed and the fans blowing on them. This along with incandescent lighting that produces over 97% heat from consumed electricity contributed to this tragedy. These efficiency effects are identical in process settings. Low efficiency increases power cost and may add heat where it cannot easily be removed.

4.20 Swamp Coolers (Evaporative Cooling)

In the South, evaporative cooling (swamp coolers) for personal use requires water and reasonably low humidity levels say less than ca 50–60%. They are effective where water is available and much cheaper than electric-powered air conditioning.

These produce solids resulting from the water evaporation process which must be periodically removed (blown down).

The industrial counterpart in the CPI and refinery makes effective use of evaporative cooling towers with similar temperature and relative humidity limitations. Costs include:

1. Water treatment, e.g., descaling and solids removal
2. Water loss make-up, impractical in water poor areas

Boilers also need high purity water and must be purged (blown-down) periodically to remove solids and sludge.

4.21 Common Equipment Efficiency Focus Points

As a note, tables of energy conversion efficiency from one form to another were presented in Chap. 3. Remember the Btu as the "coin of the realm." As part of your project or design data, start collecting this type of performance data early on. The cost of a motor can be as little as 1.7% of the total life cycle analysis (LCA) cost [8].

4.22 Energy Performance and Efficiency Consideration of Typical Chemical Process Technology Equipment

- Motors
- SE motor sets and pumps
  - VFD motors
  - Careful impeller selection and load matching
  - Electronically commutated motors (ECM)
  - Pump seal coolers in the case of accidental blinding.
  - Plant pump and piping layout
- Compressor and turbine performance curves
- HX equipment spec sheets
- Mixing performance
  - Reactors
  - Liquid extraction
- Distillation columns
  - Tray and packing efficiency
  - Throughput
  - Energy recovery
  - Reflux ratio

- Reliable voltage ±90 % and power factor >90%
- Nitrogen and oxygen onsite generation

4.23 Engineering Pilot Studies

These are often needed to assure design performance and plant operability. For new designs, these can be worth their weight in gold, and can be tied to performance guarantees. Expect to pay for these! They will help you tune your design and identify potential trouble spots including potential bottlenecks. They can also be used to reevaluate existing designs prior to cloning, so do not skimp here either! These also serve as an efficiency check point.

Pilot studies often form the basis of *Process and Equipment Guarantees* have been covered in the section on SE Technologies (Chap. 7). As noted there, this area is even more important for SE design than classic design. As SE is in its incipient stages, it could die easily if not adopted carefully and supported with fact-based efficiency and performance metrics.

4.24 Petroleum Refining Energy Consumption

Seven percent is typical overall refinery energy consumption. So, assuming 100,000 Barrel per Stream-Day (BSD) throughput, 7000 barrels is consumed as energy equivalent [9]. The production of gasoline from crude may be more on the order of 85% efficient, a fact important when compared to electric motor vehicles.

The DOE has compiled extensive equipment energy consumption–related data and information. Here are a few representative points on pumps and compressors as they have such a large SE impact over a typical lifespan of 20 years or more.

4.25 Excerpt from EPA/DOE Petroleum Refining Overview, (Chap. 3, Ref. 8)

4.25.1 Pumps

In the petroleum refining industry, about 59 % of all electricity use in motors is for pumps (Xenergy, 1998). This equals 48 % of the total electrical energy in refineries, making pumps the single largest electricity user in a refinery. Pumps are used throughout the entire plant to generate a pressure and move liquids. Studies have shown that over 20 % of the energy consumed by these systems could be saved through equipment or control system changes (Xenergy, 1998).

Correcting for pump oversizing can save 15 to 25% of electricity consumption for pumping (on average for the U.S. industry). The Chevron Refinery in Richmond, California, identified two large horsepower secondary pumps at the blending and shipping plant that were inappropriately sized for the intended use and needed throttling when in use. The 400 hp and 700 hp pump were replaced by two 200 hp pumps, and also equipped with adjustable speed drives. The energy consumption was reduced by 4.3 million kWh per year, and resulted in annual savings of $215,000 (CEC, 2001). With investments of $300,000 the payback period was 1.4 years, at 0.05 $/kW-h.

✈ [(1100 − 400/1100)] × 100 = a potential 63.7% energy cost reduction, probably more like 30% (flowrates not reported), but still this is huge. Energy audits are very useful activities. Facility managers use them to pinpoint problem areas, e.g., bottlenecks or just general improvements. ✈

4.25.2 Use Multiple Pumps

Often using multiple pumps is the most cost-effective and most energy efficient solution for varying loads, particularly in a static head-dominated system. Installing parallel systems for highly variable loads saves 10 to 50 % of the electricity consumption for pumping (on average for the U.S. industry) (Easton Consultants, 1995).

✈ This can sometimes save the cost of maintaining "hot spares" if the process can tolerate reasonable pump replacement or repair times. (Caution—OSHA Lock-out/Tag-out must be observed!) ✈

(Additional not reproduced here: Trimming Impeller, Controls, Adjustable Speed Drives, Avoid Throttling Valves, Correct Sizing of Pipes, Replace Belt Drives Dry Vacuum Pumps.)

4.25.3 Compressors and Compressed Air

A retrofit of the compressed air system of a Mobil distribution facility in Vernon (CA) led to the replacement of a compressor by a new 50 hp compressor and the repair of air leaks in the system. The annual energy savings amounted to $20,700, and investments were equal to *$23,000, leading to a payback period of just over 1 year (U.S. DOE-OIT, 2003b).*

✈ Also, of note is the National Academy of Engineering, that published the generic review of "…how the chemical industry is using energy more efficiently." ✈[10]

4.26 Pump Efficiency Example

Figure 4.2, from ITT Goulds, shows the wide range of efficiency at various impeller diameters, flow rates, and pressure loss total developed head (TDH). Here a 10-inch diameter pump, operating at 275 gpm, 100-ft TDH, and 70% efficiency consumes 10 hp. If we now reduce flowrate to 150 gpm, consumption drops to 8.25 hp. (Follow the 10-inch curve to a TDH of 112 ft.) If original TDH came 25% from elevation (25 ft) and 75% from frictional loss (75 ft), then a drop to 150 gpm reduces frictional loss to about 25 ft. New design basis is now 50 ft TDH at 150 gpm. Figure 4.2 suggests a 7-inch diameter at approximately 3.25 hp. (As theoretical is 1.9 hp, the old pump is only 23% efficient!) Increase efficiency and save even more with a different pump of a lower best operating range centering around 150 gpm. This extreme example was chosen for illustrative purposes. ITT has a downloadable design tool useful for pump selection and determining when to change rotors or pumps or both.

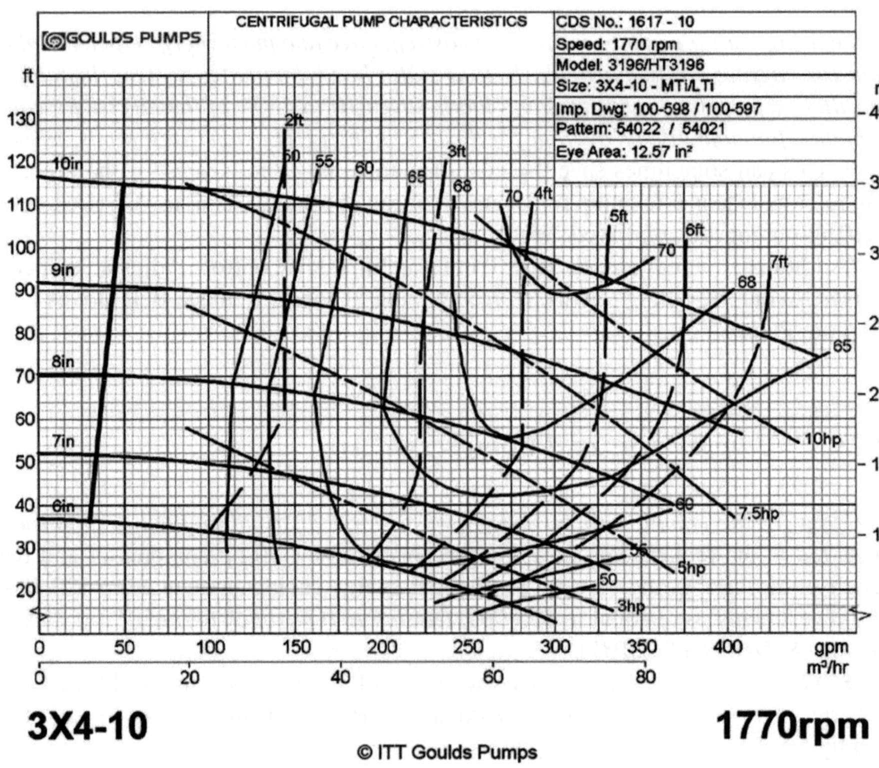

Fig. 4.2 Courtesy Rich Nardone, ITT Goulds Pumps [11]

4.27 Electric Power Challenges and Opportunities

In recent years, photovoltaics, particularly residential, have come down in price greatly. Companies have sprung up to provide free residential installation on the basis of taking cost savings for excess generation. The challenge, however, is developing a uniform method of connecting to the grid to back feed during high generation periods. My cousin in Southern California built his own small solar farm that supplied much of his power but was limited by law as to how much the grid would buy back. Still the load sharing is considerable for home and building owners.

Opportunities and Challenges include:

1. Solar, wind, and cogeneration
2. Grid interconnection
3. Power Factor management

4.27.1 Electric Grid/Production Plant Interconnection Challenges

In order to encourage medium- to large-scale on-site manufacturing power generation as described herein, other interconnection challenges must be worked out. Reactive power variation during load switching/shedding as well as power factor variations at the end user must be accommodated.

Whether for solar, wind, and cogeneration the technical issues can be worked out, but may require regulatory/financial assistance. We will look at this again in Chap. 7, Sect. 7.4.3.

Power factor (PF) refers to the efficiency of coupling power to various types of equipment. In general, the PF of direct current resistive devices approaches 100% or 1.0. For large power interconnections, this must be managed, particularly when connecting/disconnecting from the grid. These mismatches are reflected back and can cause issues.

As a final note, be careful to retain quality control over your SE design specification decisions and equipment selections. Large engineering and operating companies often have separate purchasing departments that are focused on saving initial cost in all areas, but not long-term utilities and post-construction energy consumption! The military refers to this as a sustainment tail, aka "The gift that keeps on giving." Make sure to retain signature approval on all such deviations from specification.

4.28 Summary

Control over efficiency is a key to SE success. All too often this is traded off to save initial investment upfront with little to no regard for LCA costs. If you cheat here, you earn "The gift that keeps on giving," i.e., high project life operating time cost. SE design equipment selection standards and efficiency ratings are clearly needed here.

References

1. Kyle BG (2000) Chemical and process thermodynamics, 3rd edn. Prentice Hall, Upper Saddle River
2. Callen, H (1960) Thermodynamics, 1st ed (1960) and Thermodynamics and an introduction to thermostatistics, 2nd ed (1985). Wiley, New York
3. Alternative Fuels Data Center (AFDC) Alternative fuels data. AFDC.org
4. Automobile MPG Info. Fuel economy. www.fueleconomy.gov
5. Perl JP (1989) An in-plant approach to hazardous materials management. Chemical Processing Magazine, March 1989
6. Edison T. Pearl Street Station. http://ethw.org/Pearl_Street_Station
7. Linnhoff B, Flower JR (1982) A user guide on process integration for the efficient use of energy. The Institution of Chemical Engineers, Rugby
8. Colin Coh (2013) LCA. Precicon SMC annual conference, 7 November 2013, Singapore. colinkoh@precicon.com.sg
9. O'Brien D. Jacobs Consultancy
10. Patt JJ, Banholzer WF (2009) Improving energy efficiency in the chemical industry. Energy Efficiency 39(Summer). National Academy of Engineering. www.nae.edu
11. ITT Goulds Pumps, www.gouldspumps.com
12. Kister HZ (1992) Distillation Design. McGraw-Hill
13. Lieberman NP, Lieberman ET (2003) Working Guide to Process Equipment. McGraw-Hill
14. Shinskey FG (1977) Distillation Control, McGraw-Hill

Additional Resources

Couper JR, Penney WR, Fair JR, Walas SM (2012) Chemical process equipment: selection and design, 3rd edn. Elsevier, St. Louis

Perl JP (1990) Hazardous waste treatability studies Chemical Processing Magazine., June 1990

Perl JP (1995) Technology selection and remedial design. CHMM review course, Illinois Institute of Technology

US Energy Information Agency. Basic energy utilization summary. eia.gov

American Council for an Energy Efficient Economy, www.aeee.org. Doe advanced MFC office (formerly ITP), www.energy.gov

Karassik IJ, Krutzsch WC, Fraser WH, Messina JP (eds) (1976) Pump handbook. McGraw-Hill, New York

Chapter 5
New Product Design and Alternative Process Chemistry: SE Manufacturing Choices

5.1 Bringing New Chemical Products to Market [1]

The following describes the EPA regulatory program requirements pertaining to new product development. Portions of it, however, can be used to evaluate existing chemical processes if there is interest in modifying an existing chemical with an eye toward cost and/or toxicity characteristics reduction.

5.2 The Federal Premanufacturing Notification Process (PMN) and Identification of Alternative Chemistry

Federal law requires that EPA be informed, through the PMN process, of all planned new chemical products prior to commencing manufacturing. EPA emphasis is on the "N" for notification as they do not look at this as a permit.

✈Consider this sort of a chemical's passport for entry and permanent residence into the CPI.✈

The PMN process, as described in this chapter, allows the agency to assess potential ESOH effects and to assign levels of protection during all phases of manufacturing, with special emphasis on estimated environmental emission releases. For that reason, this is an ideal point of departure for sustainability engineering design.

A result of long-term toxicology research by EPA and others has led to a scientific methodology that not only can aid in predicting human and animal toxicity but also suggest changes in chemistry structure of new products prior to production that are far less toxic while retaining the desired commercial outcomes. Companies such as 3M participated in the early evaluation of the EPA's Ecological Structure Activity Relationships (ECOSAR), a computer modeling program that continues today and is freely available to the public. As a US Air Force Reservist, I had the opportunity

J. P. Perl, *Sustainability Engineering for Enhanced Process Design and Manufacturing Profitability*, https://doi.org/10.1007/978-3-031-52363-2_5

to evaluate this program for the HQ Air Force Center for Environmental Excellence (AFCEE).

This tool, described in this chapter, is useful in the design phase for screening purposes, so make it part of your SE design procedure. EPA will use it to evaluate your submittal, so consider including it as a "toxicity simulation" tool along with incorporation into a process design package such as Aspen Plus, HYSYS, and Chemstations.

Once an understanding of the basic manufacturing process is available, the manufacturer must submit basic information including process chemistry for the agency to commence review to evaluate potential release as well as manufacturing exposure routes. The American Chemical Society (ACS) operates the chemical abstract service (CAS) and will evaluate your new chemical and provide a name and CAS number for use in the PMN application. In order to protect proprietary information, the law allows key (proprietary) portions of the submittal to be marked confidential business information (CBI) re: public access. By act of congress, EPA has a fixed, 90-day review period during which it must either approve or request further information or testing deemed necessary to complete its evaluation. Time extensions can also arise while EPA is awaiting information from the submitting manufacturer.

Various nontoxic substances, certain classes of polymers, for example, are exempt from full toxicology testing, but only with EPA approval. Other new chemicals will need to undergo ECOSAR analysis and EPA findings will dictate the type and level of lab testing, e.g., aquatic, air, and ground. These results are used to set EPA reportable quantities in the event of a spill of the new substance and can also lead to OSHA permissible exposure levels (PEL) in order to establish worker or consumer personal protection handling requirements and exposure limitations. Transportation, storage, and disposal requirements are also delineated during this stage. Again, remember this is all based on the accuracy of your plan submittals, so be diligent here and use it to tune up your process design.

The PMN, required by congressional mandate through the Toxic Substance Control Act (TSCA), is not only designed to protect the total environment but also puts review/time limitations on EPA to avoid undue hardship on industry. This is also an important part of sustainability engineering, as the ECOSAR tool can be applied to existing and planned processes to reduce toxicity and identify other safer, more sustainable feedstocks.

The sustainability engineer can review planned new chemicals well in advance using ECOSAR even before designs or EPA submittals are made. Remember, SE considerations go beyond simply ensuring renewable resources but also preventing the introduction of new, unintentional, and unwanted outcomes. Encouraging companies to assure renewable resource utilization through SE will, like automobile MPG requirements, be an economic boon for consumers as well as assuring a reduction in carbon/Btu footprint through identification and implementation of useful efficiencies.

Also worthy of note, my own PMN experience with USEPA and American Chemical Society (ACS) Chemical Abstract Service (CAS) has been entirely favorable. The staff of both entities are dedicated to protecting the public while assisting companies in working through the PMN process to its logical conclusion. The Europeans have moved ahead of us on this or at least equal, with their Registration, Evaluation, Authorization and Restriction of Chemicals program aka REACH, so this is not a process that unduly burdens US industry and hopefully will be adopted by others such as China and India. An unexpected positive consequence is the pre-design, pre-production sanity check that comes about from trying to explain to others what your manufacturing process is all about.

Here are some useful summaries of the PMN process including the ECOSAR program, from the epa.gov website [2].

5.3 Excerpts from USEPA New Chemicals Program Website at epa.gov

Your new chemical product must have a standard universally understood chemistry name, its manufacturing process must be outlined including raw materials, finished products, potential environmental releases and human exposures. Packaging, warehousing and transportation also need to be described, all to allow an understanding of the hazards and required protections for manufacturing, distribution, and end use of the new product.

5.4 New Chemicals

Mandated by section 5 of the Toxic Substances Control Act (TSCA), EPA's New Chemicals program helps manage the potential risk to human health and the environment from chemicals new to the marketplace. The program functions as a "gatekeeper" that can identify conditions, up to and including a ban on production, to be placed on the use of a new chemical before it is entered into commerce. Section 5 of TSCA requires anyone who plans to manufacture (including import) a new chemical substance for a non-exempt commercial purpose to provide EPA with notice before initiating the activity. This premanufacture notice, or PMN, must be submitted at least 90 days prior to the manufacture of the chemical and USEPA has a limit on review time to avoid undue delay/expense to production startup.

(Excerpts from USEPA Sustainable Futures Website at epa.gov)

5.5 What Is the EPA Sustainable Futures Initiative?

The Sustainable Futures (SF) Initiative is a voluntary program that encourages chemical developers to use EPA's models and methods to screen new chemicals for potential risks early in the development process. The goal is to produce safer chemicals more reliably and more quickly, saving time and money. This means getting safer chemicals into the market and in use. In some cases, it means providing alternatives to more risky chemicals—this is pollution prevention in its purest form. Companies that take training and graduate from Sustainable Futures can earn expedited review by EPA for prescreened new chemical notices. Prescreening chemicals for hazard concerns helps companies anticipate and avoid developing chemicals of concern. Companies can instead develop and commercialize safer chemicals.

(Ecological Structure Activity Relationships [ECOSAR]—New v. 1.11, June, 2012 epa.gov)

5.6 What Is ECOSAR?

The Ecological Structure Activity Relationships (ECOSAR) Class Program is a computerized predictive system that estimates aquatic toxicity. The program estimates a chemical's acute (short-term) toxicity and chronic (long-term or delayed) toxicity to aquatic organisms such as fish, aquatic invertebrates, and aquatic plants by using computerized Structure Activity Relationships (SARs).

5.7 How Does ECOSAR Work?

ECOSAR uses structure-activity relationships (SARs) to predict the aquatic toxicity of untested chemicals based on their structural similarity to chemicals for which aquatic studies are available. Application of structure activity relationships is a technique routinely used by the U.S. EPA Office of Pollution Prevention and Toxics under the New Chemicals Program to estimate the toxicity of chemicals being reviewed in response to Pre-manufacture Notices mandated under Section 5 of the Toxic Substances Control Act (TSCA). The toxicity data used to build the SARs are collected from publicly available experimental studies and confidential submissions provided to the U.S. EPA New Chemicals Program. The SARs in ECOSAR express correlations between a compound's physicochemical properties and its toxicity within specific chemical classes. Through publication of the ECOSAR Model, the U.S. EPA provides public access to the same methods the EPA uses for evaluating aquatic toxicity. Many of the SARs have been validated through studies published in the open literature or through validation activities conducted by the U.S. EPA is conjunction with other regulatory agencies. For access to some of the ECOSAR validation activities and publications, ECOSAR References section within the ECOSAR model's Help Menu, or visit the U.S. EPA's Sustainable Futures/ Publications webpage.

5.7.1 Note Regarding EPISuite and ECOSAR

ECOSAR is maintained and developed as a stand-along program. However, for users' convenience when screening chemicals, ECOSAR was included in the EPA EPISuite program many years ago so users could obtain a full environmental profile. Since that time, associated funding and maintenance schedules have become dramatically different between the two models and now the ECOSAR version in EPISuite is now out of date. We are unsure when the new version might be added.

We encourage users to try ECOSAR as a standalone program since the tool contains many additional features in the stand alone program that are not included in the EPISuite version.

5.8 Excerpt from USEPA Website Regarding ECOSAR

1. *INTRODUCTION TO THE U.S. EPA NEW CHEMICALS PROGRAM UNDER THE TOXIC SUBSTANCES CONTROL ACT (TSCA)*

The U.S. Environmental Protection Agency's (U.S. EPA's) methodology for hazard and risk assessment of new chemicals, which integrates quantitative structure-activity relationship (QSAR) models and expert systems into the hazard and exposure analysis, has been used for 30 years and reflects several specific regulatory requirements that define the framework under which the U.S. EPA must operate.

Section 5 of TSCA requires manufacturers and importers of new industrial chemicals to submit a Premanufacture Notice (PMN) to U.S. EPA/OPPT 90 days before they intend to begin manufacturing or importing a new chemical. U.S. EPA/OPPT must evaluate the chemicals for all aspects of health and safety and determine whether the substance may present an unreasonable risk of injury to human health or the environment. OPPT must make a risk-based decision on the regulatory outcome of the chemical within these 90 days. The PMN can otherwise be manufactured or imported.

In addition to this demanding 90-day review period, another constraint is that of the several hundreds of PMN chemicals submitted each year, a minority include environmental toxicity data. In response to this data-poor situation, U.S. EPA/OPPT developed "estimation methods" that are used to fill data gaps where little or no experimental measured data exist. These approaches include analogue analysis, chemical class analogy, mechanisms of toxicity, QSARs, and professional judgment. In order to quickly complete an assessment for each new chemical, the Agency uses computerized QSAR models and expert systems to make estimates for physical/chemical properties, environmental fate, ecological toxicity, human health toxicity, and chemical releases and exposures in an effort to fill data gaps (U.S. EPA 2003a). These estimates are used to support the U.S. EPA/OPPT chemical management decisions within the TSCA framework and to assist the Agency in determining the most appropriate regulatory decisions for each new chemical based on the potential risks.

This technical reference manual focuses on the scientific approach and underlying methodology for the assessment of aquatic hazards using the U.S. EPA/OPPT computerized QSAR tool called the ECOSAR (ECOlogical Structure-Activity Relationship) Class Program.

JPP—The complete text can be found at https://www.epa.gov/system/files/documents/2022-03/methodology-document-v.2.2.pdf

From the ACS Chemical Abstract Services Website, the following is provided to whet your appetite. A complete set of chemical naming instructions and contact information can be found at cas.org website.

5.9 Scientific Identification of Your New Chemical: The Starting Point for the PMN

✈ Prior to registration, all new chemicals must have a CAS number to identify it. Think of this as a social security number for chemicals that allow scientists and engineers the worldwide, to have a common, chemistry structure–based way of identification that follows accepted rules. ACS does this for a very reasonable fee. The following program excerpts explain and demystify this important sustainability program. ✈

5.10 The American Chemical Society and the Chemical Abstracts Services (CAS) [3]

Chemical Abstracts Service (www.cas.org), a division of the American Chemical Society, is the world's authority for chemical information. CAS is the only organization in the world whose objective is to find, collect and organize all publicly disclosed chemical substance information. A team of scientists worldwide curates and controls the quality of the databases, which are recognized as the most comprehensive and authoritative by organizations around the world. By combining these databases with advanced search and analysis technologies (SciFinder® and STN®), CAS delivers the most current, complete, secure and interlinked digital information environment for scientific discovery.

CAS Registry Numbers (often referred to as CAS RNs or CAS Numbers) are universally used to provide a unique, unmistakable identifier for chemical substances. A CAS Registry Number itself has no inherent chemical significance but provides an unambiguous way to identify a chemical substance or molecular structure when there are many possible systematic, generic, proprietary or trivial names.

CAS Registry Numbers are used in many other public and private databases as well as chemical inventory listings and, of course, are included in all CAS-produced databases.

✈I have added the following excerpt from EPA regarding QSAR and ECOSAR, as I believe it will be a useful SE design tool for picking and choosing alternative chemicals, or validating your original selection. ✈

5.11 Introduction to the Toxic Substances Control Act (TSCA) and the USEPA New Chemicals Program

The U.S. EPA's methodology for hazard and risk assessment of new chemicals, which integrates quantitative structure activity relationship (QSAR) models and expert systems into the hazard and exposure analysis, has been used for over 25 years and reflects several specific regulatory requirements that define the framework under which the U.S. EPA must operate.

The assessment of new industrial chemicals and the retrospective assessment of an inventory of existing chemicals are within the purview of U.S. EPA's Office of Pollution Prevention and Toxics (OPPT). The OPPT administers the Toxic Substances Control Act (TSCA) which was passed in 1976 to regulate all industrial chemicals in the U.S. Under TSCA, U.S. EPA is charged with assessing, and if necessary, regulating all phases of the life cycle of industrial chemicals including manufacturing, processing, use, and disposal. In 1979, almost 62,000 industrial chemical substances were reported to be in commerce in the U.S. and these chemicals formed the original TSCA inventory of "existing" industrial chemicals.

Chemicals not included on this original inventory before 1979 were considered "new" industrial chemicals. All new chemicals had to be submitted to U.S. EPA for review prior to commencing commercial manufacture or import activities (Zeeman et al. 1995, 1999). More than 42,000 such chemicals have been submitted by industry and assessed by OPPT since July 1979. About 20,000 of these new industrial chemicals are now in commerce, increasing the TSCA inventory to more than 82,000 chemical substances.

Section 5 of TSCA requires manufacturers and importers of new industrial chemicals to submit to EPA/OPPT a premanufacture notice (PMN) 90 days before they intend to begin manufacturing or importing a new chemical. U.S. EPA/OPPT must evaluate the chemicals for all aspects of health and safety and determine whether the substance may present an unreasonable risk of injury to human health or the environment. OPPT must make a risk-based decision on the regulatory outcome of the chemical within these 90 days. The PMN can, otherwise, be manufactured or imported.

In addition to this demanding 90-day review period, another constraint is that of the large number of PMN chemicals submitted each year (up to 2000), approximately 65% of the substances are being submitted with no experimentally measured data. Under TSCA, the notifier is not required to conduct any "new" ecological or human health testing before submitting a PMN. Only about 35% of the PMNs reviewed to date contain any type of measured data (Zeeman et al. 1995, 1999). Nonetheless, the U.S. EPA must assess each new chemical submitted regardless of the level of understanding concerning the specific chemical or chemical class. TSCA places the burden of proof on the U.S. EPA to determine whether the manufacture of a new chemical "may present" an unreasonable risk to human health or the environment. EPA cannot require the notifier to submit additional information about the new chemical unless there is an adequate basis to support an unreasonable risk finding. With this statutory limitation, and the demonstrated lack of measured data submitted with the PMNs, the U.S. EPA was faced with the need to estimate over 150 attributes for a large number of chemicals in a very short period of time in order to make rapid decisions regarding the risk associated with manufacturing a PMN chemical. Given these constraints, it was obvious that the methods of risk assessment utilized by U.S. EPA in the New Chemicals Program had to be both scientifically sound and pragmatic.

In response to this data-poor situation, U.S. EPA/OPPT developed "estimation methods" which are used to fill data gaps where little or no experimental measured data exists. These approaches include nearest analog analysis, chemical class analogy, mechanisms of toxicity, quantitative structure activity relationships (QSARs), and professional judgment. In order to quickly complete an assessment for each new chemical, the Agency now uses computerized QSAR models and expert systems to make estimates for physical/chemical properties, environmental fate, environmental toxicity, human health toxicity, and chemical releases and exposures in an effort to fill data gaps left by the PMN submitter.

(U.S.EPA2003a). These estimates are used to support the U.S. EPA/OPPT chemical management decisions within the TSCA framework and to assist the Agency in determining the most appropriate regulatory decisions for each new chemical based on the potential risks.

This technical reference manual focuses on the scientific approach and underlying methodology for the assessment of aquatic hazards using the U.S. EPA/OPPT computerized QSAR tool called the ECOSAR (ECOlogical Structure Activity Relationship) Class Program.

Here is an industrial example we worked on for the Specialty Fertilizer Company. It involved preparing the PMN as well as establishing facility costing manufacturing requirements [1].

5.12 Specialty Fertilizer Products (SFP) Case Study: Bringing New Chemicals to Market Sustainably

SFP was founded to invent and produce fertilizer-enhancement products. The company has created numerous and various nontoxic coatings and additives to increase fertilizer efficacy. We served as a consultant interface with USEPA and CAS to bring SFP successfully through the PMN process.

One of their products is a nontoxic, polymeric seed coating that helps retain fertilizer components, thereby increasing agricultural growth efficiency. Just as in refinery throughput enhancements such as distillation packing and MVR, small incremental improvements lead to large differential agricultural efficiency gains. The SFP manufacturing process uses common chemicals that are regulated by EPA and OSHA and so were well understood. A major advancement here was in the fact that the process produces inert, nontoxic polymers that protect the seed as well as the environment during the plant growth process.

5.13 Summary: "Better Chemistry for Living" [4]

Prior to the commencement of manufacturing, all new chemical products must be reviewed by EPA through the PMN submittal process. The information submitted to EPA will also serve as the boilerplate for any operating permits that will ultimately be required. The PMN process is required by law to provide an understanding of health and safety issues involved with all aspects of the new product. Use this requirement as a sanity check to review what it is exactly that you want your new chemical to do and how you will make it. What ESOH compliance programs will it need to protect the environment as well as the operating plant personnel? Is there an alternate chemistry that might be safer and/or cheaper? Since the PMN is after all a regulatory requirement, start your own preliminary PMN as soon as the team has a clear idea of potentials, and get a head start on the whole thing. And, remember, the sole purpose of this process is, amongst other things, to help set exposure and spill limits and relevant ESOH elements in order to protect the public and chemical workforce. When done properly, the process will also help in the development of all plant costs, both fixed investment and operating ones.

References

1. Perl JP, Wiggins-Lewis M, Masson RM. Bringing new chemicals to market. In: Joint meeting AIChE/CHMM, 9 Feb 2005.
2. PMN Program, USEPA New Chemicals Program. http://www.epa.gov/reviewing-new-chemicals-under-toxic-substances-control-act-tsca
3. Chemical Abstract Service (CAS) of the American Chemical Society. https://www.cas.org/
4. "Better Chemistry for Living" Copyright 1993, 2011, 2016 Jeffery P. Perl and Chicago Chem Consultants.

Additional References

Seider et al (2009) Product and process design principles, 3rd edn. Wiley, London
Cussler et al (2011) Chemical product design, 2nd edn. Cambridge Press, Cambridge

Chapter 6
Environment, Safety, and Occupational Health (ESOH) Regulations

6.1 Overview of Chemical Manufacturing–Related Federal Regulations

After publication of "Silent Spring" in 1961 by Rachel Carson, her congressional testimony, and several environmental and safety catastrophes, the ESOH issues moved front and center into the public eye.

In 1970, Republican President Richard M. Nixon signed laws establishing the US Environmental Protection Agency (EPA) [1] and the Occupational Safety and Health Administration (OSHA) [2]. These laws were extensions of limited existing ones that came to head over several large-scale environmental catastrophes. That they were signed into law by a conservative Republican President only serves to underscore the tenor of the situation during the year of the first Earth Day.

The Congress empowered EPA to regulate and establish health-based limits of toxic chemical release to the environment beyond the manufacturing plant, while OSHA was set up to watch over the chemical manufacturing workforce health and safety.

Sometimes lost in the analysis is the statutory requirement set forth by the Congress that requires all regulations to meet a cost–benefit review. The Congress set a goal to provide for the public safety, but not to bankrupt business, so an attempt to establish a reasonable balance over the last 45 years can be seen in the regulations. As more is learned about individual chemical toxicity, for instance, their allowable release and exposure levels move up or down and the regulations are adjusted accordingly.

J. P. Perl, *Sustainability Engineering for Enhanced Process Design and Manufacturing Profitability*, https://doi.org/10.1007/978-3-031-52363-2_6

6.2 SE Design Impact

About 55 years after EPA and OSHA formation, ESOH compliance has largely become second nature. What is not always appreciated however is the importance of conducting the preliminary ESOH review BEFORE principal design work commences and at least in parallel to early conceptual design. This is of particular importance when new chemicals are considered as operating permits eventually need to exist prior to the commencement of manufacturing. The whole point is to make sure a permit can be obtained first. It goes without saying that the SE material and energy sourcing review also be conducted to assure sustainability. These get staged just like all the other design elements with early broad review followed by increasing detailed ones.

Once the preliminary material and energy (M&E) balance has been conducted, a preliminary regulatory review must also be conducted. Ultimately, all manufacturing must conform to regulations and just as one would never plan a new facility that does not make a profit, a process that cannot meet regulations will never be permitted to commence operation. So, the first "0" design stage gate review should include the following.

6.3 Stage Gate "0" Preliminary Process Design Review

1. Basic Process, aka "Design Basis"
2. Block Flow Diagram (BFD)
3. M&E Source Review
4. Regulatory Review
5. Preliminary Cost Estimate

In order to create a sustainable design, SE requires that all design-related elements be optimized simultaneously. This requires drawing a larger picture around the entire system, not just ISBL. It would probably be a bit daunting to include the entire earth itself, but it must include external supply locations well outside the battery limits (OSBL) that are often on the other side of the globe. Total quality management (TQM) adherents will recognize the Quality Circle that is drawn around the "Affected Community." Businesses will recognize the need to secure a sustainable supply of raw materials. Even inside the battery limits (ISBL), and at one point a refinery flare was not considered ISBL, now it is. Personal Communication, Refinery Flares, Dennis Obrien, Jacobs Consultancy, 2015. TQM is discussed in Chap. 9.

6.4 Hierarchy of Historical Design

Old Methodology → M&E balance + profits (Pre-1970)

Design → Assure long-term sustainability regarding resources and ESOH impact (now and future).

6.5 Major Federal Chemical Manufacturing–Related Regulations

All CPI plants must comply with various ESOH regulations. Permits are issued prior to operation so all good designs begin early on with an ESOH compliance review. Prior to such regulations company's often dumped waste into adjacent rivers, lakes, and streams, out the back door, into leaking underground storage tanks, etc. This was done of course to save money and with little to no thought about health effects. As discussed before, in 1970, Federal EPA and OSHA regulations were formalized by act of congress, and signed by President Nixon making the financial burden of manufacturing ESOH regulations more even across the States. To this day, the government requires an economic burden analysis for all EPA and OSHA regulations. The government uses a cost–benefit methodology that incorporates health and other effects and therefore is dependent on science. Here are some of the larger regulations in summarized form, and by their common acronym. You can find details on all the regulations with very good descriptions and points of contacts at the websites listed. In many instances, EPA has delegated authority to the States to administer elements of their programs.

6.5.1 Clean Air Act (CAA)

The *Clean Air Act* was established to regulate the presence of contaminants that enter through the lungs or can be adsorbed onto or absorbed through the skin via exposure to contaminant-laden air. This has brought about such preventive components as the catalytic converter on cars, smokestack scrubber, and reduction of refinery vent flaring and now watching over methane releases arising from natural gas fracking. Once a material is released into the air, toxic or not, it is essentially lost to the massive diluent effect of planet earth. Keep in mind, even a low-level leak of expensive finished product can have a large effect on the economic bottom line. The same can be said for associated long-term health effects which ultimately are paid for by society. So, this is also an SE checklist item.

6.5.2 Clean Water Act (CWA)

The *Clean Water Act* was established to protect the "Navigable Waters of the USA." While RCRA tells us when spill reporting is required, CWA has special, more stringent preventive requirements designed to protect one of our most precious resources, drinking and fishing waters of the USA. Cleaning up concentrated wastes prior to entering water is orders of magnitude cheaper than after they are spilled and greatly diluted. Our entire ecosystem, and hence long-term sustainability, relies on maintaining these elements in pristine condition. Chemical manufacturing plants planning to discharge into the "Navigable Waters of the USA" must meet stringent requirements that generally discourage such discharges with their occasional accidental excesses.

6.5.3 Department of Transportation (DOT)

The *transportation of hazardous materials* falls under this department's purview. Many of our finished products are perfectly harmless, but the raw materials necessary to create them are not always so. This regulation dovetails with EPA as well as OSHA. Toxic waste and gasoline are all considered hazardous from a transportation perspective, though for different, physical safety–related reasons. Minimizing transportation of either feed or finished products is not only safer and less costly, but is also a key element of SE, thereby providing another key savings for SE-designed integrated power and production facilities.

6.5.4 Emergency Planning and Community Right to Know Act (EPCRA)

The *Emergency Planning and Community Right to Know Act of 1986* was established to allow communities to understand what hazardous materials are stored and used nearby, thereby empowering them to better prepare for any contingency. This act established local emergency planning committees (LEPC).

6.5.5 OSHA

The *Occupational Safety and Health Act* protects workers in all industries by limiting their workplace chemical exposure levels and proscribing safe physical working conditions. In the CPI, OSHA and RCRA are linked through the chemicals and hazards they both regulate, the former mostly on the inside and the latter mostly on the outside of the plant.

6.5.5.1 Occupational Chemical Exposure

OSHA regulates workplace human exposure levels of all chemicals it deems applicable and usually states this as 8-hour permissible exposure levels. Exposures in the air, significant breathing sources in particular are highlighted, as well as liquid dermal contact is included. Risk = Toxicity times Exposure, so OSHA also proscribes worker personal protective equipment (PPE) levels to control exposure in hazardous chemical environments. Some hazards lead to physiological insults, cancer, lung, and other organ disease. Some can cause fire, explosion, electrical shock, and mechanical injury. Either way OSHA watches over the overall workforce health and safety in almost all manufacturing sectors and sets permissible exposure limits (PEL) for worker protection. (FAA and Mine Safety are two that police themselves.) Keeping manufacturing processes safe is an SE design outreach. It is not just the correct thing to do for people, but worker compensation is one of the largest bottom-line expenses, either corporate or government for uninsured injured employees. In addition, insurance rates factor in incident rates so do it right the first time! Again, the USAF experience taught that protecting our workforce is not only the right thing to do, but it is also cost effective in long-term savings … a more sustainable approach to manufacturing. JPP Note—Keep that economic sustainment tail as short as possible!

6.5.5.2 Part 2 Occupational Bodily Safety

Again, Risk = Hazard times Exposure. Human endeavors often take place in the presence of hazards, such as electricity and fire explosive atmospheres, so the key to safety is to minimize or eliminate exposure.

- Physical injury prevention—Mechanical things like saw guards and the like
- Combustion and explosions—Some atmospheres are not toxic, but can explode, like natural gas
- Electrical and mechanical hazards —The Lockout/Tagout (LOTO) standard was established to prevent worker exposure to the hazardous release of energy of all types, including heat, explosive reactions, mechanical, electrical, kinetic, and potential energy

6.5.6 Pollution Prevention Act (PPA)

The *Pollution Prevention Act of 1990* is more programmatic than regulation. The EPA requires companies to state their waste minimization practices in written reports that also provide a look at the type of hazardous materials that exist within communities that could lead to accidental exposure. In the 1980s, the Congress established toxic waste land disposal bans. PPA was established by the Congress to track industry response to these bans and to make sure the toxic material was not being improperly (illegally) disposed of. This problem was so huge at the time that the Congress gave industry nearly a decade to phase in the land disposal restrictions. The PPA goals were met voluntarily and well in advance of schedule by the industry overall, with the carrot of reduced paperwork and positive publicity. Indeed, this may have contributed to the popularity of all things green as now big companies could get bragging rights about their positive environmental performance. Pollution prevention is truly a huge SE goal and one that is easily undertaken by process design engineers who now had greater leeway to include proactive P2 into their designs. The 3M Company copyrighted the phrase "Pollution Prevention Pays," because, well, it really does. Doing something correctly first leads to less waste, reduced employee exposure, and less follow on corporate liability incurred when you place your waste into other company's hands. The EPA started to focus on life cycle analysis (LCA) about this time, a way to take into account the cradle to grave cost of chemical manufacturing operations during the financial analysis phase. This is an early precursory attempt to address sustainability in the CPI.

6.5.7 RCRA

This is the principal EPA regulation covering manufacturing in the CPI. The *Resource Conservation and Recovery Act* sets allowable limits of environmental releases and includes preventive as well as corrective measures regarding spills.

✈ The Congress meant this law to both *C*onserve and *R*ecover resources making this the original sustainability legislation. Energy as well as material conservation was included. Environmental protection was, of necessity, the main focus at first.✈

6.5.8 Superfund

In the wake of several environmental catastrophes in the 1960s and 1970s, this act was brought about by companies that went bankrupt or no longer existed; the Congress set up a multibillion-dollar fund for EPA to clean up the orphan sites, and to go after perpetrators wherever possible.

6.5.9 Toxic Substances Control Act (TSCA)

The Toxic Substances Control Act gives EPA the authority to regulate extremely hazardous materials such as PCB and dioxin. It also is the window through which all manufacturing of newly developed or imported chemicals must pass prior to entering production or use in the USA. The pre-manufacturing notice (PMN) program allows EPA to identify hazards and to set permissible regulatory levels for their manufacture per RCRA. EPA also has developed tools to assist industry in identifying safer, cleaner alternatives to planned new chemicals. Interaction with OSHA occurs during the PMN process.

6.5.9.1 ECOSAR and EPISuite

These programs (Chap. 5) help *evaluate new chemicals for potential toxicity* and even show how to select chemicals with less or even no toxicity that have similar desired properties for marketable products. This is a very powerful SE design tool in and of itself to aid in avoiding poor initial decisions that have long economic lives. They also assist in preparing PMN submittals to USEPA. See also Chap. 5.

6.5.9.2 PMN

The *Pre-manufacturing Notice* program (Chap. 5) was established by EPA as a gatekeeper to entry of all new chemicals into the US market. The program is designed to evaluate toxicity to humans and wildlife and to aid in setting environmental and personal protection levels.

6.5.10 TSDF

Treatment Storage and Disposal regulations became necessary to assure proper controlled final disposition for hazardous waste. A goal of all SE is to create processes that have no need for TSDF in the first place. The Congress has been in the process of allowing certain organic type wastes to be gasified, providing a perfect SE design tool to reduce material, energy as well as CO_2 footprint.

✈ This will need to expand to all waste types if SE is to succeed as described in Chaps. 7 and 8.✈

6.6 Department of Health and Human Services

6.6.1 FDA

> *The Food and Drug Administration is the oldest comprehensive consumer protection agency in the U.S. federal government. Its origins can be traced back to the appointment of Lewis Caleb Beck in the Patent Office around 1848 to carry out chemical analyses of agricultural products, a function that the newly created Department of Agriculture inherited in 1862. Although it was not known by its present name until 1930, FDA's modern regulatory functions began with the passage of the 1906 Pure Food and Drugs Act, a law a quarter-century in the making that prohibited interstate commerce in adulterated and misbranded food and drugs.*

✈Many EPA-regulated chemicals find their way into the FDA target list. This is a particular challenge in foods originating outside the USA from countries with little to no regulation.✈

6.6.2 USDA

> *US Department of Agriculture provides leadership on food, agriculture, natural resources, rural development, nutrition, and related issues based on public policy, the best available science, and effective management. We have a vision to provide economic opportunity through innovation, helping rural America to thrive; to promote agriculture production that better nourishes Americans while also helping feed others throughout the world; and to preserve our Nation's natural resources through conservation, restored forests, improved watersheds, and healthy private working lands. Our strategic plan serves as a roadmap for the Department to help ensure we achieve our mission and implement our vision.*
>
> *The U.S. Department of Agriculture (USDA) is made up of 29 agencies and offices with nearly 100,000 employees who serve the American people at more than 4,500 locations across the country and abroad.*
>
> *Strengthen the American agricultural economy, build vibrant rural communities and secure a stronger future for the American middle class. The day-to-day operation of USDA's many programs and spearheads the $149 billion USDA budget process.*

✈Agriculture (Chap. 11) is very big business in the USA, and SE design will factor into these business segments to improve overall energy and material utilization and to enhance sustainability itself.✈

6.7 Other Manufacturing-Relevant Government Programs

6.7.1 Energy Star USEPA for Consumers

This program is identifiable by the ubiquitous yellow energy tags on appliances. The stickers, at a glance, can guide consumers to a more informed decision, without need for formal scientific training. This has been a great boon for sustainability as the market can now focus on creating not only energy-efficient appliances but reliable ones as well. The quality/reliability portion is the classic consumer decision point regarding manufacture reputation, but prior to Energy Star, there was no simple way for a consumer to purchase energy-saving devices, many of which have 10- to 20-year lives or more during which to waste energy for no reason! Always-on electronics use an enormous amount of energy with the microwave oven control center and a clock that runs 24 h using 1–3% of the annual device itself in energy consumption, drawing as much as 30 kW per year. Energy Star has gone a long way toward improvement in consumer electronic designs that go into a deeper sleep when not in use.

6.7.2 DOE Energy Programs for Industry

The US Department of Energy (DOE) [3] watches over energy reduction programs for industrial applications. Numerous technology assistance programs are run by DOE. The Energy Policy Act of 2005 outlines methods the industry and government can use to reduce energy footprints. Examples include HVAC, Seasonal Adjusted Energy Efficiency Ration (SEER), and Annual Fuel Utilization Efficiency (AFUE) regulations governing the efficiency of commercial devices. ✈ A set of guidelines identifying SE-rated equipment and methodologies might be useful, at least as a standard for industry instead of the regulatory approach taken for commercial applications.✈

6.8 Technology at Your Finger Tips—and Its Free—Well You and I Paid for It, So Use It!

EPA and DOE have a wealth of freely available information regarding regulatory compliance as well as Industrial Technology Program (ITP) reviews. Numerous engineering case studies are easily adapted for use by companies of all sizes. Technical assistance programs exist around the country, at the state, local, and federal level, that can include extremely important access to generic, nonproprietary design data so the wheel need not always be reinvented. Learn to draw upon these resources, at least as a good starting point. EPA and DOE occasionally have funds available through the small business research innovation office (SBIR). These SBA grants are offered throughout all government entities and small businesses should definitely avail themselves of these.

6.9 Example DOE Industrial Technologics Program (ITP): Summary of Program Results for CY 2009

The DOE website has a cornucopia of energy-related technology reviews. Become familiar with this free resource … free because you pay taxes!

6.9.1 Boosting the Productivity and Competitiveness of US Industry 198 Pages PDF Document

Excerpt:

> *Foreword from Boosting …*
>
> *A robust U.S. industrial sector relies on a secure and affordable energy supply. While all Americans are feeling the pinch of volatile energy prices, project financial-constriction impacts on industry are especially acute. Uncertainty over energy prices, emission regulations, and sources of financing not only hurt industrial competitiveness – together they have the potential to push U.S. manufacturing operations offshore, eliminate jobs that are the lifeline for many American families, and weaken a sector of the economy that serves as the backbone of U.S. gross domestic product.*

The Industrial Technologies Program (ITP) is actively working through public-private partnerships to address the enormous energy challenges now facing America and its industrial sector. ITP has an established track record for moving innovative technologies through commercialization and onto the floors of industrial plants, where they are at work today saving energy and reducing carbon emissions. For the period 1992-2010, ITP-sponsored projects have resulted in 50 R&D 100 awards and 265 issued patents. Also notable are the significant savings identified this year through the plant energy savings assessments conducted as part of ITP's Save Energy Now Initiative.

The daunting challenges confronting U.S. industry and the rapidly evolving energy supply situation prompted a reexamination of ITP strategies for technology development and delivery. A number of practical opportunities were identified to build on ITP strengths, expand into promising new areas, and boost program impacts to support critical national goals. ITP operates under the guidance of the U.S. Department of Energy's Office of Energy Efficiency and Renewable Energy (EERE). To learn more about the current ITP program and new directions see the ITP website at www1.eere.energy.gov/industry/

✈Boosting productivity makes US products less expensive and more competitive and amenable to domestic production, another SE positive outcome. As an example, IBM created a PC computer manufacturing operation in New York State that was nearly fully automated making domestic production economically feasible. And remember to always stick to "Fact-based decision-making" when using web-based resources. As President Ronald W Reagan said, "Trust, But Verify," and use that approach in all contracting as well!✈

6.10 ESOH Example

6.10.1 The United States Air Force Environment, Safety and Occupational Health Compliance and Management Practice Program (ESOH-CAMP)

Over 10 years as an AF Reservist, I had the priviledge to work for The Air Force Civil Engineer on environmental and pollution prevention programs. First at the HQ AF Center for Environmental Excellence (AFCEE) and later at HQ Air Education and Training command (AETC), followed by the HQ AF Surgeon General. The following example is drawn from those experiences.

The United States Air Force (AF) uses a novel evaluation program originally developed for Environmental, then expanded to all ESOH for overall compliance and management practice purposes referred to as ESOH-CAMP. Prior to ca. 1980, the military fell under the EPA radar, literally and figuratively. Then, the Congress mandated that all installations comply with the substantive requirements of EPA, usually deferring to Superfund as a model. This included cleanup of all AF bases selected by the Congress to be closed as well as those to be kept open. The HQ Air Force Center for Environmental Excellence (AFCEE) watches over cleanup and prevention at all USAF bases and installations for the Civil Engineer, USAF.

In addition, to prevent further environmental insult and promote P2, the AF created an environmental division under the AF Civil Engineer's (Major General) authority.

This became AF/CEV in mil-speak for civil engineering environmental division. This is one of the more successful compliance programs and is preventive in nature, designed specifically to assure no further ESOH degradation such as that led to the massive installation cleanup programs. If you know anything about the military, you will know that people will strive to succeed in their assigned mission as this is the path to promotion and recognition. (This is true of all human endeavors, and industry created the corporate ESOH managers for just that purpose.) And here as elsewhere, from my personal experience, the civilian and military professionals worked hard and successfully in their endeavors to improve ESOH performance service wide.

These are not classified programs and the USAF regularly has public P2 conferences to tout the successes, as well as postings in libraries local to AF Bases. Other examples include CHP to create energy, early adoption of electronic ballast fluorescent bulb re-lamping, building energy conservation, motion detection activated lighting, and HVAC improvements. Less toxic methods of electroplating and surface finishing of aircraft surfaces and other equipment were also developed through sponsored AF R&D. The AF even watched over endangered species and noise amongst the 13 environmental and 12 safety and occupational health protocols included in each assessment visit.

These methods are entirely exportable to civilian use. A biennial review process had our bases competing to outperform each other. This featured annual self-inspections, followed by HQ ESOH-CAMP team visits to evaluate performance. This allowed HQ to develop a metrics-based methodology for allocating pollution prevention program funds to where an actual net positive rate of return could be had. At one point, the savings had totaled nearly $400 million.

As an example at one such AF installation a novel recycling program collected all manner of waste plastics, glass, metals, and paper. All were separated appropriately and stored for sale only when the market pricing was high. This program generated well over $150,000 annually and paid for staff as well as avoided placing this material in MSW landfills. And take note—this methodology is applicable to industry, but is often given little attention by busy production-oriented staff. The developers and operators of the program at this AF base received an award, given by the Base Commander as well as additional funds they requested to enhance paper collection on this base, home to 10,000+ military and their families and several thousand daily personnel. ESOH-CAMP is an extremely successful AF-wide program saving over $400 million through proactive ESOH compliance and improved operational efficiency.

6.11 Presidential Executive Orders

Presidents have a long history of issuing PEOs [4] often politically contentious. They have been used for everything from national park creation to establishing environmental and energy efficiency requirements for government facilities, i.e., everything from office buildings to military installations. Most have survived presidencies due to their utility from one party to another. ✈ https://www.federalregister.gov/presidential-documents/executive-orders/ ✈

6.12 Summary

So why is this section in this book? ESOH should always be considered prior to committing to even a preliminary design effort. Do not work on something that will be too unsafe to be permitted. The review process is surprisingly uncomplicated, generally leading to safer and more profitable manufacturing alternatives and their attendant designs.

References

1. EPA—US Environmental Protection Agency. www.epa.gov
2. OSHA—US Department of Labor, Occupational Safety and Health Administration. www.osha.gov
3. DOE—US Department of Energy. www.doe.gov
4. Presidential Executive Orders, The White House. Whitehouse.gov

Chapter 7
ChE SE Technology Equipment and Utilization Toolbox

7.1 Sustainability Engineering Technical Additions to Classic Design

Let us keep in mind that SE really is just an incremental albeit important improvement over existing process design, and product design and development engineering. As such, the classic methodologies of economic evaluation, product development, conceptual as well as detailed design engineering, process control, maintainability, operability, profitability, etc. all still hold. The big differences of course include:

1. Renewable, continuous supply of materials
2. Renewable, continuous supply of energy
3. Integrate power generation and manufacturing process when justified/possible:
   (a) Generate H_2, NH_3, or others with off-peak electricity—the "Battery Concept," for use later here or by others
   (b) Small and large scale
4. Minimal to zero adverse effect on ESOH
5. Tighten up efficiency, maintenance, operability, plant operations, and profitability
6. KEY—Extend SE material and energy balance timeline beyond desired plant life → sustainable

Classic investment grade engineering economic analysis is challenging at best, even when cloning existing plants. Sustainability analysis requires an even greater depth of technical analysis or it may not be sensitive enough to point to the optimal solution. For example, Chap. 4 is devoted to the efficiency of everything. By this, it means the difference between theoretical performance, say, of rated pump horsepower and actual shaft horsepower coupled into moving the fluid. Add to this the

J. P. Perl, *Sustainability Engineering for Enhanced Process Design and Manufacturing Profitability*, https://doi.org/10.1007/978-3-031-52363-2_7

challenge of integrating disparate plant processes, and typical variability of pump curve efficiency, and it is easy to see how multiplicative efficiencies can greatly affect the SE determination more than in the past. The incorrect pump size, while meeting flow, pressure demand, and initial cost requirements, can easily use up to 50% more energy … forever! A pump curve example illustrating characteristics including efficiency is shown in Chap. 4, Fig. 4.2.

7.2 Sustainability Engineering Definition/Criteria: Key SE Principle

The timeline for assured resource availability and integrated components must extend beyond the expected or required subject operating plant life to attain project sustainability. Simply put, if a project has no net impact on resource availability or ESOH, then there would be no measured difference in availability, between time $t = 0$ and $t = 20$ for those resources, for a 20-year project or process life, then that project would be considered sustainable. Note this has to include any post-operations waste generated and system decommissioning, such as nuclear or other power plants. You cannot simply walk away at the end of year twenty and forget the closure costs.

Green way of looking at it—you want to leave things pretty much the way they were at first. Even more simply, sustainable means leaving the entire planet the way it was before you started your local project, wherever that may be. (Naturally, improvement is always acceptable!)

Business in general, is not very good at looking beyond a 2–3 year ROI for investment purposes. There are many variables in play such as war, economic downturns, acquisitions and divestitures, failure to comply with the manufacturing laws, e.g., ESOH, between countries to name a few. NAFTA is a good example of such a failure. As a reminder: US law prohibits the export of waste to NAFTA countries, but there is no mechanism to enforce EPA- or OSHA-type transgressions outside of the USA, leaving compliant US manufacturers at a competitive disadvantage. Hopefully, the new TPP will address this SE deficiency as it too puts compliant US businesses at a competitive disadvantage with noncompliant trade partners. [In January 2017, the President withdrew the US from the TPP. Later that year, the other countries created their own TPP so I have left this reference from the first edition (2016) as it may yet reappear.]

7.3 The Btu as the Coin of the Realm for Sustainability: A Key SE Parameter

Sustainability engineering analysis, particularly with regard to comparing options, must rest upon a stable, conservation law of physics–based methodology. So, options that cause Btu footprint increase are inferior to those with reduction. In a perfect world, money and energy would be interchangeable, but that is not the world we live in. So energy conservation must be the scales of sustainability justice. Similarly, process options consuming greater materials must also be considered less sustainable than those requiring less. SE requires the overall material and energy utilization footprint be reduced or held constant in the process.

As previously mentioned, I helped a client prove a significant Btu footprint reduction for a proposed process manufacturing change affecting their commercial product sold by their customer, WalMart, a major US retailer. This is much more involved than merely adding up dollars on both sides of the equations, but when it is done on an energy basis, it provides a permanent physics and fact-based methodology and not one that merely looks good at the time it was presented. So here, the Btu was the coin of the realm and an 86% energy savings was made. No surprise that this also saved the manufacturer by avoiding the cost in chemical solvents as well as unnecessary processing steps. Again, another unintended consequence turned positive. And this is typically the case, i.e., improving the process almost always leads to cost savings. The sustainability "birthing process" is painful, but will lead to long-term benefits.

7.4 SE Elements to Coordinate Plant Wide

- Material manipulations
- Energy manipulations
- Onsite power production
- Integrate all three for full SE
- Always in ESOH compliance

7.4.1 Material Manipulation: It All Has to Balance

At steady state, the plant-wide material balance must be maintained. What goes in must come out, however much it may have changed appearance. The method of creating finished organic (merchant) products however is selectable. For some products, there is little to no variation in allowable feedstock. When looking to enhance the SE performance, gasification is a versatile tool.

7.4.1.1 Gasification: The Premier SE Tool

For better or worse, society as well as industry produces enormous amounts of waste. But this waste is a definition relevant to the process itself. Per EPA regulations, a material is not a waste until you say it is, and adding a gasifier onto the back end of a process unit, or as a material and energy generator upfront, is a SE design modification for existing plants or the right way to do it for future plants. Pyrolysis may also be used where applicable, but may include oxygen as chemically necessary, increasing air emissions. Of course, technically both gasifiers and pyrolysers are just two types of chemical reactors.

As noted in Chap. 3, gasification of almost anything organic into CO and H_2 can provide a superior feedstock to replace or augment crude oil and coal. Greater conversion and selectivity can be expected with far fewer side reactions that require expensive purification steps. And if you make energy along the way, that's SE! There is no reason for any organic waste material to go unused and gasification is the one-stop tool to make this happen.

In the past, there was an enormous pushback against incineration of all types of wastes. The fear of exhaust air emission gases containing toxic elements was real and experience based. Not in my backyard (NIMBY) was the hue and cry of communities. In the future, communities will want this ultralow emission SE tool to help reduce carbon footprint, generate otherwise lost energy, reduce or eliminate MSW landfills completely, while generating local revenue.

Gasification can be used either to transform waste into products, or create feedstock from high value organics, e.g., natural gas, oil, or even coal, or both.

7.4.2 Energy Manipulation: Double-Entry Balance with Materials

Of course, minimizing plant energy use is key to the success of any operation. But with SE methods described here that might include power generation, a new level of sophistication is both required and rewarded at the same time.

7.4.2.1 Heat Exchanger Networks (HEN): Moving Energy from Point A to B Within a Plant

Distillation columns and reactors are used extensively within the CPI. Individual heat exchangers are used to move waste heat from one stream to another. For complex processes, there can be dozens of these HX, AKA economizers. It is useful in SE to design them as a network, to optimize their use and avoid thermal crossing, AKA pinching (Chap. 4, Sect. 4.7). Figure 7.2 shows a distillation column before and after addition of both heat pump and feed HX economizer. An example of divided wall distillation provides a novel internal HX scheme with no moving parts, the details of which are presented at the end of this chapter.

7.4.2.2 Heat Pumps: The Energy Fulcrum

As described in the Chap. 4, the mechanical heat pump is a remarkably simple device for raising the thermal potential energy level of vapor streams. Referring to Fig. 7.2 at the end of this chapter, when used in distillation, the compressor takes the place of condenser, cooler, and boiler. The overhead vapors of the column are compressed just enough to raise its temperature sufficiently above the boiling point of the bottom of the column or any other adjacent column. By raising the thermal potential energy of the vapor, the heat pump provides the bottom reboiler heat duty, taking load off the boiler, or in most instances, removing the boiler completely. Typical tradeoff is temperature differential versus cost of equipment purchase and operations. Economics generally favors low delta T.

The hot vapors are introduced to the reboiler where they condense into hot liquid. This hot liquid, which is actually the distillate product supplies reflux with the remainder contacted against a cold stream, typically a feed, to recover even more heat prior to product discharge. This "low-quality heat" cannot be directly employed in a heat engine, and this is the one of the great benefits of minimizing energy consumption in CPI applications, regular as well as integrated. This is part of the HEN described in Sect. 7.4.2.1.

The concept of coefficient of performance (COP) is essentially the same as in HVAC equipment and can even be higher, as much as 10/1 owing to the use of both high- and low-quality "waste" heat in the manufacturing processes. The tradeoff, of course, is the relatively high cost of turbo-compression machinery required for these operations as well as added maintenance requirements. But the savings obtained by avoiding purchase of condenser, cooler, and boiler, along with the COP-based energy savings should lead to paybacks of less than 3–4 years, with 20+ year life and on the order of 75% CO_2 reduction.

So, the SE designer must learn how to incorporate this economic decision-making and to demand high-efficiency equipment. The use of continuously variable high efficiency motors must also be employed and this takes practice, and good contracting! I also like the energy efficiency ratio (EER), the mechanicals used to describe air conditioning efficiency. EER = Btu Produced/Watt-Hour consumed. In the CPI the goal of integrated processing is also to increase the numerator. ✈ This can also be done in remote jungle areas where free wood waste is available for fuel. Power and potable water can be made. (J. Perl, EWB 2010 Chicago) [1] ✈

The heat pump and its big brother the heat recovery steam generator (HRSG) are a part of SE energy conservation either in standalone or integrated disparate chemical process production. A heat pump BFD is shown at the end of this chapter. The HRSG is covered in greater detail in Sect. 7.4.3.1.

7.4.2.3 Process Energy and Steam: Back Together Again for the First Time

The modern petroleum refinery is a good place to look for integrated process energy. Many exothermic reactions generate steam used elsewhere. Power is not commonly generated, but the refinery does a great job in general with the material and energy balancing. Onsite power production is a key SE item.

7.4.3 Onsite Power Production

This is a key link for SE success. Creation of steam for process use is commonly done in boilers. A special, high-efficiency system, borrowed from the power industry, is described in the next section.

7.4.3.1 Heat Recovery Steam Generator (HRSG) Electric Power Generation: A "Plugin" SE Power Source

This technology is commonly employed in modern electricity generation plants to create steam that is continually reused at different pressures to increase internal efficiency and then for "low-quality" heating purposes. Thomas Edison recognized early on the enormous heat wasted in his first electricity generating stations and set about to capture and reuse this in what became known as district heating.

As natural gas production through fracking has become more prevalent, this can allow the CPI to generate electricity for self-sustaining plant use and to capture an enormous percentage of waste power production heat, both high and low quality, for plant use. The natural gas as we will see in some examples can also be used as a chemical building block through gasification to create useful merchant products such as ammonia for fertilizers and other chemicals. This, however, requires locating disparate facilities together to best take advantage of this in pure SE mode.

This might include, e.g., locating desalination plants alongside highly exothermic process technologies that produce otherwise unusable waste heat, a perfect match for the energy-intensive distillation and evaporation required for salt and dissolved solids removal. Here, water purification and production would be the integrated element with power. Not so farfetched for a CPI that considers water along with power as a utility.

By itself, the HRSG is only 40% efficient, so move your combined heat and power (CHP) generating plant together with CPI to make a totally integrated power and production facility. Here are the usual plant process design considerations.

7.4.3.2 Companies That Build and Service Onsite Electricity Generation Systems

I found examples of integration of disparate chemical processes to optimize energy and materials utilization, but not with onsite generation of power for internal use and/or external sale. Here are, a few companies, however, that build and service onsite manufacturing facility electric generation systems:

1. Drax focuses on helping companies in going off the grid
2. GEVernova
3. Siemens Energy
4. NAES—power management and power generation services

Excerpt from EIA on factory energy generation:

Most industries purchase electricity from electric utilities or independent power producers. In addition, some industrial facilities also generate electricity for their own use using fuels that they purchase and/or the residues from their industrial processes. For example, many paper mills have combined heat and power plants that may burn purchased natural gas or coal and black liquor produced in their mills for process heat and to generate electricity. Some manufacturers produce electricity with solar photovoltaic systems located on their properties. Some industrial facilities sell some of the electricity that they generate.

Some manufacturers produce electricity with solar photovoltaic systems located on their properties. Some industrial facilities sell some of the electricity that they generate. Industry uses fossil fuels and renewable energy sources for: Heat in industrial processes and space heating in buildings

7.4.4 System Integration of Process Materials and Energy and Power for Maximum SE

This is where SE really shines. Once the primary plant design basis is set, an additional new effort to locate the appropriate mix of power and disparate processes will be needed, but the payout can be large. In the future, this step will foster new, profitable alliances in the CPI and related industries.

7.4.4.1 Material Integration with Onsite and Offsite Distribution

Several examples will feature gasification in the remainder of the text. In the case of natural gas rich regions, excess gas can be taken in to feed process gasifier(s), HRSG for plant power and provide excess surge capacity to surrounding industry and/or homes. Oil and MSW can take the place of natural gas.

7.4.4.2 Power Integration and Production for Onsite and Offsite Distribution

Once the optimal collection of disparate processes becomes the design basis, a review of onsite power consumption needs is prepared. This includes all electrical and steam needs. Then, an initial HRSG can be specified. The classical material and energy balance will be affected by the availability of high- and low-quality HRSG heat, a positive SE aspect.

7.4.4.3 Plant-Wide Combining Elements: A Few Common SE Design Process, Utility, and Offsite Needs

In classic design mode, the better-designed plants will typically integrate heat loads from reactors, columns, chillers, and perhaps cooling towers. With SE power integration, the process material and energy needs, utility, and offsite demands are tuned with power to extract maximum energy while using minimum materials. With tight SE planning, an additional feature can be zero or near zero discharge of non-merchant product streams. There is little to no need for waste to move off the site with SE.

Some classic design elements modeled somewhat after the proprietary UOP-Honeywell process design methodology include:

Process Equipment Needs

- Heat exchangers
- Columns: distillation and evaporation, scrubbers, liquid extraction
- Reactors
- Pumps
- Compressors
- Piping
- Controls

Plant Utility Needs

- Water
- Electricity
- Fuel
- Steam

Offsites

- Power generation plant, boilers, cooling towers, and air separation plants
- New—Integrate power with process, offsites, and utilities to attain true SE

7.5 Some Generic SE Tools for Technology Examples

The name of the game here is adherence to SE design requirements. Paying attention to renewable resources, efficient equipment, and integrated power opportunities. Once the SE design basis is established, the good news is that most of the equipment you will specify will be commercial off-the-shelf technology (COTS). But watch out for poor or sharply peaked equipment efficiency versus capacity. Standards will be needed for SE and we will look at this in Sect. 7.12.

I have added here a few of the usual suspects. Some of these are available now, some require R&D polishing. Tools do not produce merchant products per se, but are rather drop in elements for SE-based chemical process production. The two categories are physical and chemical processes. Here are just a few SE-applicable equipment and reformatting tools.

7.5.1 Sample Physical Operations Tools in the CPI

- Mechanical vapor recompression
  - Four- to fivefold heat energy advantage for *distillation* and *evaporation*
  - Recover maximum process materials
- Hybrid electric generation—CPI style
  - Utility two-way connection
  - Process integration
  - Recover maximum process energy
- Geothermal energy source and sink
  - Leverage large energy reservoir of the earth

7.5.2 Sample Chemical Reformatting Tools

- Gasification
- Pyrolysis
- Traditional reactor systems
- MSW mining, organic, and metals with huge positive environmental and cost impact

7.6 Some SE Tool Descriptions Expanded View

Just a sampling here for demonstrative purposes, of some SE technologies for use in SE process design.

7.6.1 Algae to Oil: A Material Resource and CO_2 Sink

Major challenge today is to apply ChE principles to miniaturize this natural photosynthetic process. In particular, development of bioprocess acceleration methods to reduce by at least three orders of magnitude space required. These might be coupled with solar collection systems that recover heat or photovoltaics. So a continuous algae-to-oil reactor might one day serve as a drop-in fuel supply replacing petroleum or natural gas, or as a material feed supply. Waste digestion as well as animal digestion time frame might give a clue to further development.

7.6.2 Bio-methane Gas Production: An Energy Resource

Methane through fracking is becoming quite popular and will hopefully continue development. But watch out for issues such as subsidence, which also plagues agricultures' excessive fresh water well draw-down.

Fracking is an important addition to the energy mix along with petroleum, but in the long run does not meet strict sustainability renewable resource requirements. It may qualify, however, as a nonrenewable extension and bring the SE rating into usable range creating a bridge to the future.

7.6.3 Municipal Solid Waste Processing: Renewable Process Resource of the Future

Cities might be best positioned to take advantage of economies of scale to collect methane gas from garbage in the numerous MSW landfills, but any collection entity can do this. In MSW landfills, energy and materials may be recovered simultaneously just as they can in a properly designed chemical production facility … pretty good when the feedstock is free, such as MSW! Note: someday MSW might be sold as a commodity feedstock when folks wake up to its value. This might need further regulatory development, and certainly more vision than most city government's exhibit. Collocating CPI facilities near these facilities can lead to SE economies of scale. In most communities, segregated recycles, e.g., plastic, glass, and paper are picked up for free with the local community government benefiting. This also avoids landfill disposal costs and can create yet another feedstock stream for SE-designed CPI facilities.

7.6.4 Contaminated Soil Remediation: A Material and Energy Resource

Numerous leaking underground storage tank contamination sites exist around the country. Also, at one time before natural gas became the city street lighting choice, town gas was manufactured by pyrolysis of coal and wood. This created a medium Btu fuel used for street lighting and cooking, but left behind a toxic, tarry substance that was buried on the site in pits. These pits are still being cleaned up by excavating and removal for offsite treatment or disposal. This incredibly wasteful process came about due to the pushback against onsite incineration. This was driven by local fears of incomplete combustion leading to toxic emissions. No consideration was given, however, to the far safer process of gasification, which can create H_2 and CO for use either as a clean fuel onsite, or to manufacture other products. Onsite gasification of organics should be examined for any site contaminated with hydrocarbons to avoid the huge cost, and hazards of transportation of such material. This material should be converted into syngas. Not buried!

7.7 Water Consumption and Treatment: A Perfect Power and Process Integration Partner

When it comes to SE, water is a very special case all unto itself. Some processes consume water in reactions or blending recipes; some facilities merely use copious quantities for cooling. Some such as agriculture place large irrigation demands on the ever-decreasing aquifers or reservoirs. Whatever your case, SE demands special attention to water as it once was taken for granted as a relatively small utility cost. As a simple example cities are just waking up to the need to reduce toilet bowl flush quantity.

7.7.1 Potable Water: Conserving and Keeping It Clean

Potable drinking water is one of our most precious and ever more expensive resources. Industry commonly consumes copious amounts either as chemical reagents in the recipe (cannot save much here) to tank washing, and conductive and evaporative cooling, where savings through SE design can be quite large. Pumping groundwater beyond rain recharge rates is clearly unsustainable. Treated wastewater should *not* be arbitrarily placed in rivers, streams, or the ocean unless justified by downstream use. Such water can be better used. These designs must allow for long time ROI to assure proper SE methods are employed upfront instead of requiring expensive revamps later. Remember, some of the SE optimizations may have to have a longer economic evaluation window. Much like a child, they may need protection to get them to maturity, but afterwards will reap benefits.

7.7.2 Desalination: The Perfect Waste Energy Sink and Integrated Power Partner

The world's supply of potable water is rapidly disappearing and the means to remove salt and other impurities is costly and often energy intensive. Virtually all industry needs process water and those situated in coastal areas can benefit by incorporating waste process heat to sustainably purify and produce their own water. Parts of California are literally experiencing massive subsidence as groundwater is pumped down at ever increasing and insatiable quantities. This is unsustainable and agri-business would benefit from SE-integrated production.

The oceans are a near perfect receptor and holder of atmospheric recycled water … not from treated wastewater, but from rain! Living by Lake Michigan has taught that even the massive great lakes are not as large as the ocean and so annual levels are too dependent on where the rain falls in the Midwest.

Desalination [2] of sea water is extremely expensive through classical means, e.g., distillation, even with MVR and multiple effect evaporation (MEV). Solar can help a bit but integrating electric power with heat generating chemical production can provide a good deal of clean water for free by linking with exothermic processes. This is a perfect SE Integrated approach application.

✈Keep in mind, as global warming increases, regardless of cause, glaciers and ice cap melting will continue to cause sea level rise, requiring engineering design or costly existing plant modifications. ✈

Some of the SE drop-in technologies to recover potable water are as follows.

7.7.3 Water Treatment Technologies

Some of the water treatment technologies include:

1. Membranes
2. Multiple effect evaporators
3. Heat pumps (MVR)
4. Dividing wall distillation
5. Process heat recovery from highly exothermic reactions

Water is a commodity far too precious to waste and there are many ways to avoid including:

7.7.4 Reuse Treatment Plant Wastewater

1. Grey water—use it on lawns
2. Cooling water—does not need to be potable
3. City Water—also need not be potable

7.7.5 Wastewater Reuse: Just Like the Astronauts

Here in Chicago, we are finally getting around to sterilizing our treated wastewater that is returned to local rivers. This water was already extremely clean and amenable to natural sterilization by friendly native water-borne species bacteria over medium lengths of time, but still can lead to localized issues where human contact is concerned, e.g., swimming. Already, the Chicago River has become alive once more with fish. But instead of dumping this treated wastewater back into the river, why not add it back into the drinking water mix, or at least use in irrigation as suggested recently by the mayor of Los Angeles? Take care, this might create a perfect localized solution, but if the receiving river is actually using this water downstream then all bets are off. This is another example of SE design integration that will require a strong team approach with an interesting quality circle! These groups will need to learn how to work together and not on opposing teams for the betterment of society as well as the good old bottom line. Here the bottom line is cost savings to tax payers by turning a waste into a feed. The SE economic analysis is no different here than that used to decide to build a manufacturing facility.

✈Chicago recently started charging more for drinking water which had been sold for much less than its value and in some instances, given away for free. That which is free is not valued and typically wasted. The same is true of garbage collection, which now will be charged for. What does this mean to SE? Very simply this offers new, more concentrated MSW feedstocks to make recycling more profitable, as well as providing an additional incentive for community recycling of glass, metal, paper and plastic. ✈

7.7.6 Grey Water—Lawn Sprinkling: A USAF Experience

The USAF Environment, Safety and Occupational Health Compliance and Management Practices Assessment Program (ESOH–CAMP) has been very successful in designing and monitoring the use of grey water, e.g., nontoxic rinsate like dishwater, for lawn sprinkling. This is extremely effective in the warm southwest and southeast where many of our flying training bases are located, saving millions of gallons of water per year. Also of interest, when I was a wee lad, there used to be something called "City Water"—nonpotable, cheaper, and less wasteful for industrial use only. Here are some other examples, ripe for SE improvement:

- Cooling water—Industrial: does not need to be potable
- Rainwater collection—Urban: good for lawns
- Rooftop collection systems—Use green roofs for heating, cooling, and gardening

7.7.7 *Water Filtration and Purification*

Filters have long been used to purify liquid streams of solid contaminants, or to recover useful solids from those liquids. They can sometimes replace or augment more expensive options, e.g., distillation and evaporation. Here are a few examples.

7.7.7.1 Membrane and Other Filtration Processes

- Simple fabric, diatomaceous earth, porous media—for solids removal
- Granular activated carbon
- Ion exchange resins
- Micro-filtration—for small particulate removal
- Ultra-filtration—for size exclusion of particles and some large molecules
- Reverse osmosis—for ultrapure water generation

7.7.7.2 Water Purification

- Granular activated carbon
- Ion exchange resins

7.8 A Few SE Process Production Tools and Considerations

7.8.1 *Fluid Plant Pumping: The Forgotten Energy Sink*

A considerable amount of energy is consumed in fluid plants moving material around. Of course, some of this is necessary, but SE design will require application of high-efficiency pumps, correct pipe diameter and material and optimal routing to accommodate both operability and maintenance while also minimizing pump energy consumption. This is sort of the stepchild of process and does not always get the attention it needs. You only get one chance to get this right upfront to avoid "the gift that keeps on giving," i.e., permanent high pumping cost. Do not forget the proper selection of high efficiency motors as well. Remember large quantity of small, distributed energy losses can lead to large costs. Consider using variable frequency or equivalent drives in place of valve arrangement to save power costs, and watch the efficiency!

7.8.2 Differential Contacting for Tank Cleaning to Conserve Water or Solvent

One of the odd, but practical things I learned in school was this concept. My lab experiments often involved the study of the surface tension and activity of various liquids and solids and surfactants, and this required ultra-clean glassware. The EPA has a lab glassware cleaning criterion that might start with soap, and various acids, but ends with triple rinsing (as opposed to filling and draining). From mass transfer concepts, spraying the side of the glass is far more efficient and effective than filling and draining a tank completely. This is not intuitively obvious and this practice should be incorporated into all designs where such tank cleaning is required. When water was plentiful and cheap, this was not as necessary, but this practice of spray rinsing is an important sustainability enhancement. Spray nozzles are industrially available for these applications.

7.8.3 Nitrogen Scrubbing of Solvents to Recover 99% + Solvent with Water and Distillation

Here is an enhanced solvent (methanol) recovery method using a recirculating nitrogen blanket system vent scheme to meet process explosion proof requirements as well as greatly increasing solvent recovery for economic purposes while staying well below environmental permitted release requirements—a win-win for SE.

In Fig. 7.1, nitrogen is used to blanket vessels containing flammable/explosive solvents. The tank vents are sent to the scrubber where water is used to recover the solvent. By removing non-condensable nitrogen, the alcohol can be concentrated from this aqueous solution by simple fractional distillation and then sent for process reuse. In principle, this method can recover nearly 100% alcohol and greatly reduces air emissions, in theory to near zero, and the N_2 is recycled as well. Add MVR to the distillation column to enhance SE! (I had to convince the client that the high N_2 recirculation rate was not the consumption rate!)

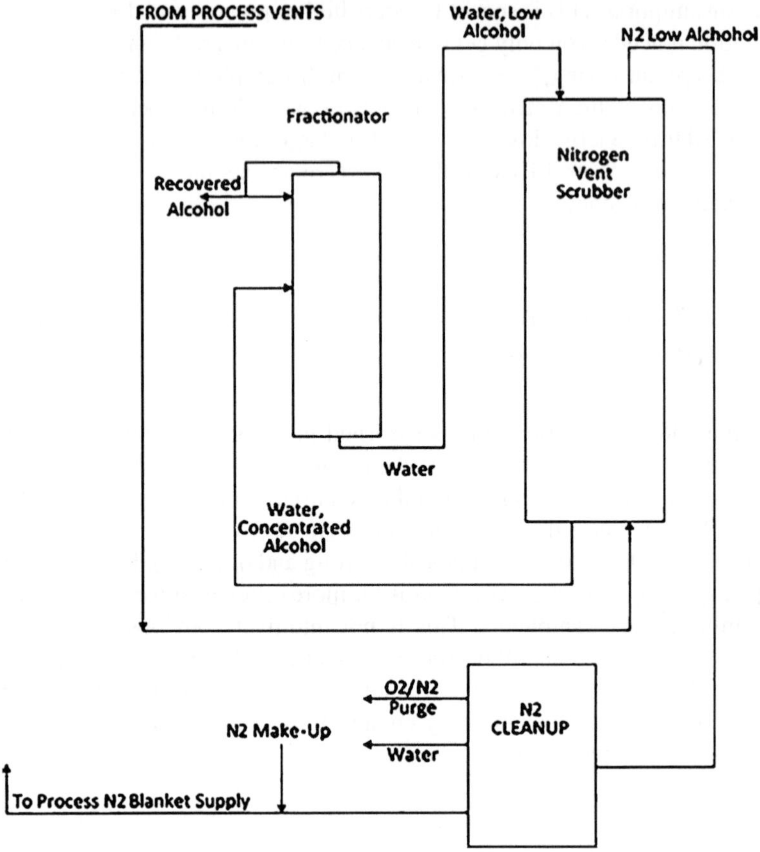

Fig. 7.1 Alcohol nitrogen process vent scrubber and recovery

7.8.4 Process Vent Condensing Vapors in the Presence of Non-condensable Gases

Thousands of process exhaust vent condensers exist that employ very expensive cryogenic or once-through cooling water systems for vapor air emission control and recovery. These ultra-low temperatures are required to literally freeze out solvents which, in the presence of noncondensable vapors such as nitrogen, are not amenable to standard condensation. The scrubbing system as shown in Fig. 7.1 easily concentrates organics into a high recovery efficiency distillation column with an overall low energy footprint. But some facilities are hobbled by outdated permitting requirements that need to be made more flexible to accept such SE process improvement changes. JPP Note: I have run up against manufacturers who are not willing to visit with EPA regarding these changes for fear of _______ you fill in the blank! My experience with EPA has been largely positive, particularly when you are offering a pollution prevention improvement. And remember, when it is your job to oversee pollution prevention you will work hard at it and EPA will typically always support this. Again these changes almost always lead to long-term cost savings and in some instances, reducing emissions can reduce or eliminate reporting requirements. Again, Congress established EPA to protect the public, not bankrupt industry. There is a reason they regularly give out pollution prevention awards. Find the appropriate folks at EPA, make them part of the quality team, and move on.

7.9 Energy Storage

7.9.1 Elevated Water Storage: Your Own Mini Hydroelectric Project at a Fraction of the Cost

If land is available, pumping large quantities of water to 100–300 ft elevation during the night to access off-peak power plant electricity rates can be employed. Works like this; pump up cheaply at night to reap off-peak electric rates then flow down against generators, saving on-peak rate as well as transmission cost. These can also be used to store solar or wind power as well. Water, tanks, and pumps are relatively inexpensive to purchase and maintain, but large amount of real estate is required [3, 4].

7.9.2 Off-Peak Electricity Storage with Ammonia

Ammonia generation has been proposed from excess electricity, as a vehicle fuel via hydrogen generation [5].

7.9.3 Using the Grid with Integrated Power Generation

If you make power onsite, the grid becomes your offsite surge tank, a sort of battery, not so much as an actual surge, but more as a control element. Here the process is king. You float your excess power on the grid, and can even buy when grid prices are low, e.g., off-peak if that works! Also during off-peak hours when the grid might not be able to accept power, electricity can be used to split water into H_2 for later fuel use and O_2 for breathing air, welding, or wastewater treatment. This is a good example of the level of planning and design sophistication required for SE that if incorporated can lead to improved overall economics.

7.10 Material Storage

7.10.1 Concept of a Sustainability Surge (Material Storage) Tank: New Application of a Tried and True Process Methodology

If nonstorable energy or material becomes available, use it and take sustainability credit. Do not let the over-ripe banana go unused … make banana bread! Surge tanks are a necessary design component of any chemical process as you cannot typically go directly from one unit operation into another without some sort of flow/ concentration buffering. The same is true with energy and mass. Flywheels already do this, e.g., hybrid cars recharge the battery during breaking and deceleration. Find economically viable methods to account for this material and energy other than throwing away. For example, ammonia can be produced from excess electricity, especially cheaper at night (off-peak), and then stored for later conversion to hydrogen for fuel use [5].

7.11 SE Economic Considerations

All projects must go through a proposal, then evaluation, then to a Go/No-Go decision process. Identifying renewable resources is, however, somewhat new. A longer supply life is more sustainable.

1. *Life cycle assessment*—Assess ecological burdens and human health impacts around entire process system [6]. ✈LCA is an early precursor to sustainability engineering that looks at "cradle to grave" disposition of materials. LCA can be used as part of the sustainability overall economic review. ✈
2. *Traditional net present value* (NPV) and internal rate of return (IRR) evaluations still hold true of course, but the project lifetime should be less than the resource availability for SE to hold.

7.11.1 Process and Equipment Performance Guarantees

In general, designers are aware that all process equipment has an efficiency of less than 100%. The SE design engineer must become schooled in selecting high-efficiency equipment as well as demanding performance up-time and longevity guarantees. I have seen examples where overly optimistic performance expectations led to process and economic failure. All parties must build-in realistic, defensible contractual requirements that include total plant commissioning, to determine actual installed performance as well as plant startup to assure operability and maintainability, none of which are obvious from written representations.

The proper upfront use of expensive energy recovery and reuse equipment can greatly enhance project sustainability. The careless application of such technology however can lead to mistrust of such approaches and a shying away for future applications. Keep the snake oil salesmen away and cultivate the good ones as well as good SE practices! Poorly written contracts can lead to expensive change orders. While there is nothing new here, mistakes will reflect poorly on SE methodology and we do not need scapegoats.

7.11.2 Equipment and Systems Commissioning and Testing

As mentioned in the section on the efficiency of everything (Chap. 4), such guarantees will need to be tighter than ever before if sustainability engineering is to establish and maintain a foothold in the design world. Cost overruns will not be tolerated here as they are in the traditional EPC. Keeping vendors and clients tied together right from design basis development through plant commissioning and handover is a recipe for long-term success and growth for both parties. This was true before and more so for SE.

7.11.3 Enhanced SE System Performance Contracting and Evaluation

Process equipment should always be purchased with a performance-based guarantee. This has been a contract requirement for classic design by engineers and procurement experts probably ever since the second plant they built! Such agreements serve to clearly state and delineate the design basis, overall equipment duty as well as other design features. So, for a pump or compressor, this would include, among other things, inlet and outlet pressure delta P, total developed head, TDH, liquid density, specific gravity, boiling and freezing point, viscosity, explosive (XP) versus nonexplosive atmosphere, and pH over expected range of use. The pump vendor will then select a pump, motor, and impeller combination unless you specify a preference.

Each motor, pump, and impeller will come with a set of performance curves. These are graphical representations of efficiency as a function of common parameters such as rpm, flow rate, temperature, pressure drop, and horsepower. The SE design engineer, along with the client, has the opportunity to specify equipment that will operate most efficiently over the expected range of conditions. And this is pretty much what is done during an initial ideal preliminary design process. A sample pump efficiency curve is shown in chapter [5].

So far so good for a simple noninteracting process, but for SE with full power and multiple process integration, a greater level of attention to detail will be required. All processes must come under automated control. In the case of complex, highly integrated SE requiring some form of real-time interactive modeling, spreadsheet, or computer, e.g., Aspen, ChemCad, or others will be necessary to dampen out positive feedback oscillations that can cause severe damage. The efficiency of all components, therefore, will need to be known in continuous digital format, and should be as high as possible and as flat as possible to avoid such process upset disturbances. This is analogous to avoiding the incorrect state in a multiple steady state chemical reactor.

Modern equipment under electronic controls is easily programmed to shutdown to prevent damage, out-of-specification product, or to avoid or prevent dangerous operating conditions. Compressor surge algorithms are a good example. Fractional distillation columns having narrow acceptable composition ranges, reactors with specific conversion and selectivity requirements or having multiple steady states offer just a few more examples. Of course, for a single train process, control under these conditions typically present resolvable design challenges, but when multiple processes are linked, say, to steam electric generation, let down energy taps, column pump-arounds, and other unrelated, disparate processes that all must be kept operating safely and optimally, then a deeper level of control will be needed for successful SE plant operation.

The deeper level of control for such SE designs is also more critical in terms of meeting profitability, as it is embedded into the very definition of SE.

7.11.4 Sustainable Process Construction Contracting Checklist

- Design and construction
- Startup
- Commissioning
- Training
- Handoff
- Remain in the loop throughout contract terms

7.11.5 Example: Post-construction Estimate Difference—Commissioning Versus Design

The examples in this book, for integrated power and chemical production are based on the newly developed natural gas supplies coming from fracking operations. The following item, taken from Chemical and Engineering News (C&EN), discusses estimates of theoretical versus actual methane release during the fracking process.

Not surprisingly, the theoretical design loss estimates were well below field experience. Here are some recent technical headlines regarding this from C&E News and the EPA response.

7.11.5.1 Leaks During Methane Production Underestimated: C&E News, September 14, 2015

*The potent greenhouse gas methane is leaking at higher-than-expected rates from a largely unstudied part of natural gas operations—facilities that collect, compress, and process natural gas for pipeline distribution. (Environ. Sci. Technol. 2015, DOI:*https://doi.org/10.1021/acs.est.5bo2275*).*

A new study—the first national one on methane emissions from such facilities—found that these operations had methane losses twice that of earlier estimates by the Environmental Protection Agency. Anthony J. Marchese of Colorado State University and colleagues sampled ambient methane concentrations at 114 gathering facilities and 16 processing plants in 13 states and used computer methods to extrapolate their measurements to calculate nationwide losses. The new figures raise estimates of the total methane loss during oil and gas production from 1.3 % to more than 10 % of final production yields.

EPA estimates that if methane leakage exceeds about 3%, the climate benefit from natural gas over coal is lost. The amount of gas lost from gathering facilities is valued at $390 million a year, Marchese says, and could provide enough gas to fuel 3.2 million households.

➔It is not clear who will pay for this. Perhaps the fracking companies will choose to treat these costs as acceptable. No doubt profits are large now, but not many production facilities would be happy with a 250% increase in energy loss. Watch those process guarantees and make sure they are performance based. They will undoubtedly require a project insurance carve out. These losses are clearly a result of non-SE design work. ➔

Here is the recent potential fracking emissions response from the EPA.

7.11.5.2 New EPA Rules Would Cut Methane Emissions from Oil and Natural Gas Industries (By Krishnadev Calamur, August 18, 2015, GovExec.com)

The Environment Protection Agency is proposing a new rule that would reduce methane emissions from oil and natural-gas drilling by 40 to 45 percent of 2012 levels by 2025. The rules would also amend existing regulations and be applicable throughout the oil and natural gas industry, including in production, processing, transmission and storage, the EPA said.

➔Most of SE will not have such interaction with government, but more likely with the CEO or corporate board. Bottom line is do your homework regarding process estimating, be they classic- or SE-based. The EPA changes described here are quite likely to happen as the estimated impact on global warming is quite large. Such large losses really should support recovery efforts, if not just for safety sake! Sound economics and good engineering, in the SE way, should prevail. ✈

7.11.6 Economic Dislocations

Economic dislocations will, however, exist that could confuse and upset the balance of things, but remember, there is no law of conservation of money. Think of the manner in which gold varies in price. (And no, NPV determination does not cut it here for the very same reason!) For example, the rapid natural gas price drop from 10 → 3 $/million Btu, or petroleum from 100–30$/Bbl might save money in the short term, but does not directly save energy and may even encourage waste. Neither does cutting down foreign rainforests for wood at 1/20th the cost of domestic wood, a completely unsustainable activity without some sort of countervailing husbandry activity. So, $$ are not always a good measure of sustainability. Use the sustainability index (Sect. 7.12.2), or Table 7.1. This is one great challenge of SE, i.e., how to balance the $ versus Btu equation.

7.12 SE Standards Development: The Next Big Thing

The Btu footprint of SE technology is important. Being able to conveniently select such equipment for use in SE designs will greatly aid in such efforts. The overall SE design itself should be distinguishable from its non-SE counterpart in some

Table 7.1 Proposed sustainability index ranking—key point

• UNSUSTAINABLE 1–10 year life, no sustainability features
Existing, non-SE design methodology
Resource depletion assumed
• GOOD 10–15 years
Fully integrated materials and energy and power
Minimal to no adverse impact over project life
• BETTER 15–25 years
Fully integrated materials and energy and power
No adverse impact over project life
• BEST 25–50 years
Fully integrated materials and energy and power
No adverse impact over project life
• GOLDEN 50–100 years
Fully integrated materials and energy and power
No adverse impact over project life

quantifiable manner. In the future, business loans, marketing programs, and consumer interest might make good use of this information. Government business might require it. More work needs to be done, so here is a point of departure for such quantification efforts:

7.12.1 Sustainable Technology Certification

This should be linked to efficiency as well as interoperability, or the ability to operate between disparate processes. This could be similar to Energy Star for industrial process technologies. It might make the selection for use in SE-designed processes easier by weeding out unacceptable technologies. It might also lead to a renewed, healthy competition among process equipment vendors to supply the "best for the least."

7.12.2 Sustainability Engineering Design Certification

Here is a proposed sustainability engineering design rating system that might make a good point of departure for rating SE design projects. This is for nontraditional, i.e., not oil and gas applications. Availability of these two commodities is ever changing as exploration does continue to find more proven reserves, but most are in areas requiring recovery from seabed or via fracking. Another major blowout like the one BP experienced in Macondo well off the Louisiana Coast, or a major despoliation of groundwater associated with fracking, might severely limit those reserves ability to contribute to the "proven" category. A similar pushback against nuclear electric generation occurred after Japan's Fukushima Daiichi Reactor meltdown of March 2011. The same phenomenon happened in the USA in the 1970s as a result of the Three Mile Island reactor meltdown of March 1979.

The primary objective here is for SE design to serve as a bridge to the future. By aiming even for good SE rating, the likelihood of bridging into a new era is quite likely. The concept of 100 years would, by all current evaluation, be considered infinity. That is, if we can get a technical approach designed that will have no adverse effects for a period of 50–100 years, then we are free to work on future, next generation, improvements. I doubt that anyone could have predicted 3D printing 10 years ago, nor could anyone predict what its ultimate impact may be on manufacturing yet.

I note that GE claims a sustainability focus for its ecomagination program. Clearly, there needs to be a way to vet technology. Vendors already have standard process guarantees and warranties, so it remains the responsibility of the SE designer to properly specify design basis, efficiency, throughput, energy consumption, and demonstrated power curves for turndown scenarios. Here is a proposed good, better, best sustainability rating schema that is based on the life expectancy of a resource. I note that even the sun will burn out someday!

Pay careful attention to the equipment you purchase today with an eye to overall project life. Designer and client must team up on this as client will pay dearly and for the life of the project for inefficient technology. As noted already, the use of integrated power and processing designs will require even higher efficiency as well as tighter efficiency versus load performance in order to assure process interoperability, maintainability, and of course, profitability, always safely and sustainably.

7.12.3 The Need for Careful Review of Sustainability Criterion

Forcing SE for SE's sake is not the best approach. An attorney wins in court whenever they uncover people lying, and no matter how small the lie, this seriously undermines their credibility. The same goes for SE in that it must always withstand the test of fact-based science and sound engineering principles, or risk political demise where public and or regulatory support may be needed.

JPP Note—Just look at the state of affairs surrounding scientific evidence regarding global warming or for that matter, anything where such evidence might fall on different sides of political arguments. The Daubert rule in trial court would be useful here. Daubert makes it illegal to lie scientifically, or to knowingly use junk science. While successful defense against junk science can sometimes be difficult, however, it is rarely due to dual interpretations of fact. Someone once said "you are entitled to your own opinions, but not to your own facts." Perjury of any kind, technical or otherwise, is against the law!

Once again, reducing carbon footprint is a natural result of SE design practice which comes about as a result of conservation both of material and energy resources and greater reliance on renewable ones, regardless of philosophy.

7.13 Detailed Example: Heat Pump in Process Application

Figure 7.2 shows the generic application of a process heat pump as applied to classic fractional distillation. In the MVR system shown at the bottom of the Fig. 7.2, the compressor serves to elevate the thermal potential energy of the overhead distillate vapors for reuse in the bottom of the still. A COP approaching 5:1 can easily be had, minimizing energy consumption by 80%. The purchase of the compressor is offset by condenser, cooler, and boiler which are not needed as well as the energy savings. This application is limited to relatively low top-to-bottom column thermal differential.

There are also other ways to use heat pumps in processes, including inter-process heat transfer in HENS.

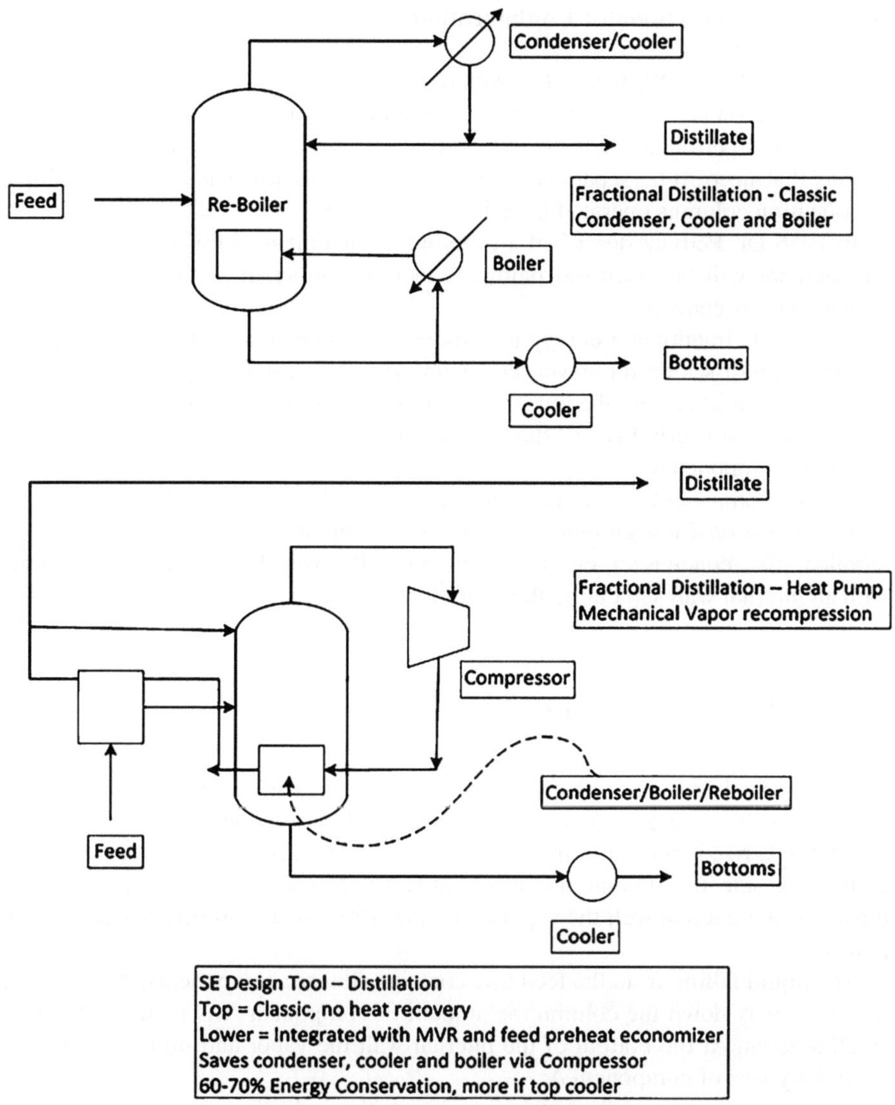

Fig. 7.2 Heat pump—distillation process classic versus integrated energy recovery

7.14 Contributed Item: Divided Wall Distillation

Here is a review of a novel energy saver regarding fractionation that is not all that well known or understood. Thanks for the following item goes to my colleague from Jacobs, who participated in development work on dividing wall during his 34 years of refinery design and troubleshooting at UOP. As presented here, this tool can be plugged into any process with multiple fractionation needs.

Dividing Wall Fractionator Configuration
Contributed by

Dennis O'Brien, PE, Jacobs Consultancy, Chicago IL

Fractionation is a very widely used separation technique in the chemical, petrochemical and petroleum industry. Many fractionator feeds contain multiple compounds that need to be separated. For about 100 years this has required multiple fractionation columns operated in series.

In 1956 Dr. Petlyuk described a different arrangement of this equipment. This arrangement will be discussed below. It took a number of years for industry to implement this concept.

In a simple fractionator energy is introduced at the bottom of the tower. Vapor is generated and rises up the tower, contacting liquid that is falling down the tower. Part of this liquid is from the tower feed and part is from reflux liquid from the top of the tower. At every tray or "stage" in a packed tower vapor and liquid mix and exchange components.

If three components are to be separated with high recovery and high purity for each, conventional design requires two towers. This means two tower shells, two reboilers, two condensers, etc. In a dividing wall tower design one tower is used with an internal wall separating the tower into two parts.

7.14.1 Simple Dividing Wall Description

On the feed side the light component (here named A) and the mid boiling component (B), and the heavy component C are separated. A plus some B and a very small amount of C rise in the vapor from the feed tray. As this composition makes its way up the feed side, the amount of component C is reduced to a very small amount. At the top of the internal wall the vapor and liquid streams are essentially free of component C.

The liquid falling from the feed tray contains all three components. As the liquid makes its way down the column the amount of component A is reduced to a very small amount. At the bottom of the internal wall the vapor and liquid streams are essentially free of component A.

The lowest section of the tower (i.e., below the wall) the separation of components B and C is performed. Component C is removed out the bottom and component B is removed as a sidecut draw on the product side of the dividing wall.

The upper most section of the tower performs a similar separation of components A and B. Component A is removed overhead and component B is removed in the sidecut draw.

7.14.2 Dividing Wall Advantages

With appropriate margins for feed variations, product purity requirements, etc., the dividing wall can perform a better separation with less energy, less employed capital, and less maintenance than the two-column configuration.

In many implementations, the dividing wall tower is 15–30% lower in capital and 30 % lower in utilities. These are very obvious savings.

There are other savings that should be taken into account:

1. Less process plot area—savings in land use
2. Less storm run-off to be treated—lower quantities of wastewater to processing
3. Less area and volume requiring lighting
4. Less area that must be in a fire protection area
5. Less equipment and therefore lower maintenance costs

7.14.3 Some Users of Dividing Wall

BASF has been very innovative in the use of dividing wall fractionators. These fractionators provide a significant competitive advantage for the company.

In the last 10 years a number of refining companies and process licensors have implemented this new equipment arrangement:

- SASOL in South Africa has the world's largest dividing wall fractionators.
- ExxonMobil has converted several towers to dividing wall to perform the benzene, toluene, and xylene separation.
- UOP has designed and built a number of dividing wall towers in units licensed around the world. These include applications in detergent plants, and reformer and naphtha cracker feed processing units.
- Valero has built four reformer-feed dividing wall towers to reduce benzene precursors.

7.14.4 References for O'Brien: Dividing Wall

1. BTX Fractionation Conventional, Pressure Cascade or Dividing Wall, Laura Weaver, Dennis O'Brien, AIChE Spring Mtg. 2011, Chicago IL
2. Tutorial on Dividing Wall Columns, Doug Stewart, Mike Schultz, Dennis O'Brien, Spring Mtg. AIChE 2001
3. Reduce Costs with Dividing-Wall Columns, Mike Schultz, Doug Stewart, Jim Harris, Steve Rosenblum, Mohammed Shakur, Dennis O'Brien; CEP Magazine May 2002, PP64-71. (Extensive bibliography)

7.15 Summary

This chapter presents an overview of practices needed in conjunction with classical design and economics methodologies to support SE design. In this chapter, we started with technology examples, e.g., equipment, and, in Chap. 8, we will examine cobbling them together in complete SE systems.

References

1. Perl JP (2010, 31 Aug) Energy and material sustainable resource management. In: Paper presented to engineers without borders, Chicago
2. El-Dessouky HT, Ettouney HM (2002) Fundamentals of salt water desalination. Elsevier
3. How do pumped-storage hydro plants work? www.duke-energy.com/about-energy/generating-electricity/pumped-storage-how.asp
4. Tom Murphy (http://physics.ucsd.edu/do-the-math/2011/11/pump-up-the-storage/) for a humorous, albeit physics-based look at pumped energy storage
5. Evans B. Convert excess electricity into NH3 for later H2 fuel. Space Propulsion Group, 24 September 2013, www.spg-corp.com
6. Life Cycle Assessment, Additional Summary this website. www.PE-International.com

Additional General Resources for this Chapter

Baker RW (2004) Membrane technology and applications, 2nd edn. Wiley
Bloch HP (2006) A practical guide to compressor technology, 2nd edn. Wiley-Interscience
Cheryan M (1986) Ultrafiltration handbook. Technomic Publishing Company
Couper JR, Penney WR, Fair JR, Walas SM (2012) Chemical process equipment—selection and design, 3rd edn. Elsevier
Das T (2005) Towards zero discharge. Wiley
Doherty MF, Malone MF (2001) Conceptual design of distillation systems. McGraw Hill (Note this could also be in ch 3)
Kidnay AJ, Parrish WR (2006) Fundamentals of natural gas processing. Taylor and Francis CRC
Lieberman N, Lieberman E (2003) Working guide to process equipment, 2nd edn. McGraw-Hill
Rushton A, Ward AS, Holdrich RG (2000) Solid-liquid filtration and separation technology, 2nd edn. Wiley-VCH
Cabezas H (2013) Design of sustainable energy supply chains using the P-graph methodology employing multiple metric Critera, AIchE Process development symposium 2013, Oak Brook, June 2013
Kister HZ (1992) Distillation design. McGraw-Hill

Chapter 8
SE Industrial Process Examples

8.1 Some Sustainability Project Examples: A Broader Perspective

8.1.1 Manufacturing Scale Approach to Material and Energy Optimization for Sustainability Engineering Design

JPP Note: Actual CPI plants, employing combined disparate processes where reaction byproducts from one merchant product are processed in another as feed, are rare and probably proprietary. One example from Black and Veach, which does not include power generation and export, is in this text (Fig. 8.5).

I have also placed two combined, interconnected process examples that include power generation in Appendix C. These were created by my colleagues and I to teach my senior chemical engineering design students about this highly synergistic approach. Power integration and distribution of electricity OSBL is also key to energy minimization. One, in particular, was located in Williston ND and took Fracking gas to use in the chemical process, while generating sufficient electricity for plant operations as well as export OSBL to the surrounding community of electricity and natural gas.

Since the first edition of this book, the USDOE has begun focusing on what they call onsite energy programs that, while not identical, do capture the synergy of such operations, also sometimes referred to as combined heat and power (CHP). The website also lists funding opportunities USDOE has to promote this approach. The overall aim is to reduce greenhouse gases along with significant concomitant energy and production cost savings. (Note: Thomas Edison recognized CHP in his early power generation stations (Ch. 4, Ref. [6]).

The E in USDOE stands for energy, and this is a serious mainstay of the Department. They sponsor and publish a wealth of information relevant to production energy efficiency and improvement, very relevant to SE as described here.—JPP.

J. P. Perl, *Sustainability Engineering for Enhanced Process Design and Manufacturing Profitability*, https://doi.org/10.1007/978-3-031-52363-2_8

We will follow with examples from USDOE, AIChE, my client files (sanitized of course), and two special cases posed by my colleagues and I for use in teaching my course in chemical engineering process design.

8.1.2 Onsite Energy Excerpts from USDOE

The excerpts are taken from https://www.energy.gov/eere/iedo/onsite-energy-program (Onsite Energy Program).

8.1.2.1 What Is the (USDOE) Onsite Energy Program?

The Onsite Energy Program is a new initiative to provide technical assistance for industrial facilities and other large energy users to increase the adoption of onsite clean energy technologies.

Technologies include but are not limited to:

1. *Battery storage.*
2. *Combined heat and power (CHP).*
3. *District energy.*
4. *Fuel cells.*
5. *Geothermal.*
6. *Industrial heat pumps.*
7. *Renewable fuels.*
8. *Solar photovoltaics.*
9. *Solar thermal.*
10. *Thermal storage.*
11. *Wind power.*

8.1.2.2 Why Is Onsite Energy Important?

By generating and storing electricity and heat directly at their own facilities, manufacturers can save money, reduce uncertainty associated with fuel prices, and gain greater control over the availability of clean energy and how it gets integrated into their processes. Many onsite energy technologies save energy and reduce operating costs by increasing efficiency and capturing usable energy that would otherwise be wasted.

Developing clean energy resources onsite can also help decarbonize industry. Clean onsite energy technologies can provide facility owners across the industrial sector with a practical option to reduce their emissions and dependence on fossil fuels, by generating electricity and heat from flexible, reliable, and affordable energy resources.

8.1.3 Manufacturing Efficiency Excerpts from USDOE

The excerpts are taken from https://www.energy.gov/eere/iedo/energy-efficiency-technologies.

8.1.3.1 Energy Efficiency Technologies

Industrial Efficiency & Decarbonization Office.

8.1.3.2 What Is Industrial Energy Efficiency?

Innovative energy efficient industrial technologies require less energy to perform the same or similar function as current technologies.

A particularly important opportunity to improve energy efficiency is in developing technologies to recover, store, and/or use waste heat. In 2018, 12 quadrillion British thermal units ("quads") of thermal energy were used onsite in the manufacturing sector in 2018, with 7 quads of total energy lost as waste, according to a Manufacturing Energy and Carbon Footprint analysis. While energy losses cannot be brought to zero, limiting losses and reducing final energy demand both offer pathways to decarbonize manufacturing and reduce costs. Ref https://www.energy.gov/eere/iedo/manufacturing-energy-and-carbon-footprints-2018-mecs

8.1.3.3 Why Is RD&D in Industrial Energy Efficiency Important?

In 2020, the U.S. industrial sector accounted for about 1/3 of the nation's primary energy use and energy-related CO_2 emissions. Without intervention, industrial sector energy demand could grow 30% by 2050—increasing CO_2 emissions by 17%, according to data from the Energy Information Agency. This is why lowering primary energy demand is key to achieving short-term emissions targets and achieving economy-wide net zero emissions by 2050.

Adopting energy efficient industrial technologies allows manufacturers to:

- *Reduce their carbon footprints.*
- *Lower costs, raise productivity, and improve shareholder value.*
- *Improve performance.*
- *Meet environmental standards.*
- *Create energy efficient products and market opportunities.*
- *Improve their competitive position.*
- *Attract top talent looking to work for a company aligned with their values.*

Ensure better community relations and an overall better reputation with consumers.

8.1.4 IEDO Research in Industrial Energy Efficiency

IEDO supports the research, development, and demonstration (RD&D) of technologies that reduce primary energy demand and minimize losses.

Technology areas pursued by IEDO include:

- *Process intensification to combine separate unit operations into a single technology, resulting in a more efficient, cleaner, and more economical manufacturing process.*
- *Energy efficient technologies with applications in multiple industries that use low-carbon fuels and energy sources or clean electricity.*
- *Low- or no-heat process technologies which achieve similar end products to current processes while utilizing significantly less thermal energy, such as mechanical separations or electrochemical processes.*
- *Flexible combined heat and power (CHP), which generates electricity and heat on-site and allows the ratio of electricity to heat to be varied as needed. Ref https://www.energy.gov/eere/iedo/combined-heat-and-power-chp-and-district-energy*

Industrial heat pumps, which can upgrade waste heat to useful temperatures using a relatively small amount of energy, ultimately resulting in net energy savings.

Next, we will look at some small industrial examples without power integration. With the advent of usable micro-turbines, it may be possible to generate plant electricity on a small scale from onsite waste products or other feedstocks. Furthermore, solar projects might also be coupled in here as well. Several large-scale examples with power integration are presented later that were used in teaching integrated power and process design in the senior chemical engineering process design course I taught over a period of 6 years. I have also included some examples of published current sponsored research to support manufacturing improvements.

8.1.5 AIChE Literature Dive

My professional organization, AIChE, is an excellent source for SE tool box items. These are portions (tools) of technologies that might be applied in aggregate to enhance SE. For brevity, I have included brief descriptions along with links. These are just little snippets, so please check out aiche.org for the complete source.

8.1.5.1 AIChE Chemical Engineering Progress (CEP) Magazine Snippets (Edited portions of the following examples were taken from CEP Magazine):

8.1.5.2 Catalyzing Commercialization September 2023

Catalyzing Commercialization: Novel Treatment Process Makes Fertilizer from Human Waste (https://www.aiche.org/sites/default/files/cep/20230913.pdf).

This article in CEP discussed methods for waste water plants at capacity to meet nutrient removal regulations. To save on cost of sewer upgrades, NSF sponsored research methods are described.

This technology development was supported by the NSF Small Business Innovation Research (SBIR) program.

This article was prepared by the National Science Foundation in partnership with CEP.

**.

8.1.5.3 AIChE CEP Special Section: The Energy Transition—Energy Update

Special Section July 2023 Chemical and Engineering Progress Magazine.

https://www.aiche.org/resources/publications/cep/2023/july/special-section-energy-transition-energy-update

8.1.5.3.1 Process Heating: A Key Step in Industrial Electrification

The chemical process industries (CPI) are consume large quantities of fossil-fuel in powered operations, e.g., *process heating ... electric heat sources are starting to replace fossil fuel combustion in plants and factories.*

1. *Process heating—for drying, melting, and cracking processes—accounts for 45% of energy consumption in industry. Companies like Shell and Dow are moving to reduce fossil fuel footprint and to save money, on industrial heating.*
2. *Steam cracking is being electrified a collaborative development agreement to accelerate technology for electrifying steam cracking furnaces. The collaboration, which began in 2021, is part of both companies' commitments to becoming net-zero-emissions businesses by 2050.*

In July 2022, Shell and Dow's experimental e-cracking furnace was successfully started up on Shell's Energy Transition Campus Amsterdam in The Netherlands.

8.1.5.3.2 Decarbonizing Heat with Long-Duration Energy Storage

(LDES). LDES technologies, which can store energy from 10 hours up to days or weeks, address the intermittent nature of renewables like solar and wind. https://www.ldescouncil.com/ldes-technologies/

8.2 Small, Nonpower Integrated Stand-Alone Process Examples

Some examples from my client files as well as others is followed by full SE combined chemical manufacturing and power generation for onsite consumption and export to the (OSBL) surrounding community.

8.2.1 Example 1: Cleanup of Contaminated Soils

This was a contaminated used oil dump site that was cleaned up under voluntary action between industry and government. The contaminated soil contained:

- 20% oil
- 10% water
- 70% dirt and clay
- <1% sulfuric acid, trace chlorinated compounds, and PCB.

A classic approach to this problem is to dig up material and haul to offsite incinerator for combustion, typically with no recovery of heat or material. A total lost cost.

An alternate SE approach involved onsite excavation, evaporation, and condensation of oil for recovery for reuse by oil recycler. Cleaned soils were returned to their original location.

8.2.1.1 SE Technical Evaluation Review of Example 1

To make a process like this work better, the regulations regarding waste definitions and ultimate treatment need to be addressed nationally. Substances are toxic only through exposure. There are many ways to process organic materials in such a manner as to avoid unnecessary detoxification and move directly to reincorporation as a useful feedstock. The oil mentioned in this example had low levels of chlorinated substances and was an ideal candidate to supplement crude oil in a refinery with little to no extra processing.

This takes load off the petroleum refinery 's raw material side and, in turn, the energy required to produce the oil, while having the additional benefit of also reducing the load and handling of the so-called toxic wastes. This will require simple reconfiguration of existing laws to assure the prevention of what EPA refers to as "sham recycling," but has great potential of also reducing carbon footprint as this material is often burned with no commensurate benefit. By definition then, this process would also be a CO_2 footprint reducer.

Refineries typically pushback against adding material of varying consistency into the crude oil processing stream. Of additional concern, present EPA rules could render an entire refinery "Toxic by Definition," if even a small amount of contaminant were accidentally introduced, providing an example of regulatory changes necessary to encourage SE design/behavior. Here, one could consider the gasifier as a tool to convert such oil into CO and H_2 for use in the field or in the refinery for chemical production, e.g., ethanol (Schultz, Michael, LanzaTech Process Description, Personal Corp Communication). Within a refinery the gasifier can be operated in either surge or production mode. I refer to this process as chemical reformatting and the refinery is an ideal place to do this, but it is also amenable to onsite processing, e.g., gasification and creation of H_2 and CO for production into other useful chemicals, e.g., NH3, again reducing CO_2 footprint at the same time.

8.2.2 Example 2: Locomotive Rebuilding Degreasing, Cleaning, and Oil Recovery

Rebuilding railroad locomotive diesel engines requires cleaning and removing sludge, dirt, and oils prior to machining operations. Early methods involved soaking and spraying parts in chlorinated solvents. These solvents were later found to be toxic and proved difficult to control or prevent accidental releases, leading to worker and environmental exposure.

Alternative engine rebuilding preparation involved exposure to hot alkaline aqueous sprays in an enclosed booth. The spent water/sludge/oil required separation and destruction of materials. A method of oil separation employing ultrafiltration was used allowing extension of spray bath water from 1–2 weeks up to 3 months. We demonstrated major savings using ultrafiltration (UF). Here, the UF membranes employed allowed passage of soap and water, while rejecting oil. This results in two useful streams, thereby reducing both water consumption and expensive detergent consumption, while providing recyclable oil for reuse.

Figure 8.1 shows the process prior to adding ultrafiltration—this engine-degreasing hot alkaline part washer consumed expensive detergent, with bath water changed weekly at best and sometimes more often. Engines are placed in the spray washer cabinet and cleaned of oil which is drained along with water and detergent and sent to treatment. This featured a high loss of expensive "designer" detergents and low concentration oil which limited feasibility of recovery.

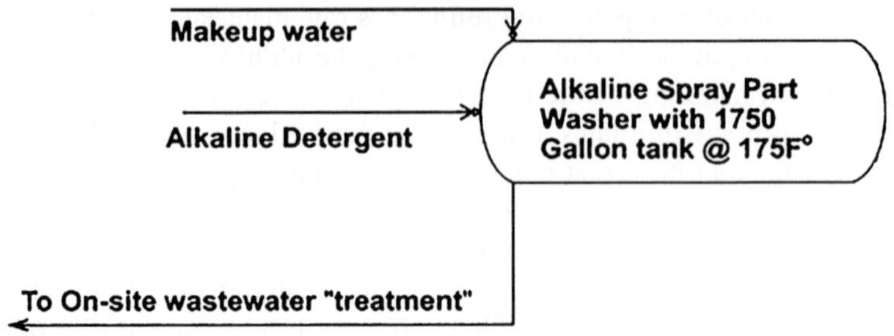

Fig. 8.1 Diesel locomotive engine part washer—original

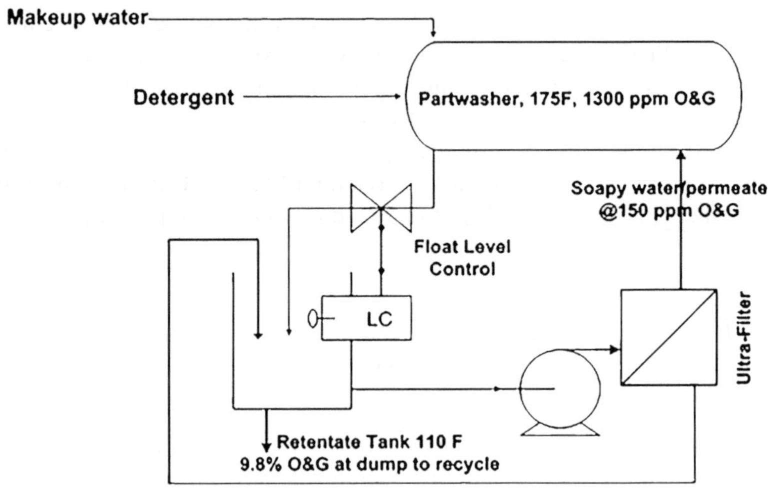

Fig. 8.2 After addition of ultrafiltration

Figure 8.2 shows the process after the addition of ultrafiltration (UF) system: Following degreasing, oil, water, and detergent are sent to a UF molecular size exclusion process that passes water and most detergent while excluding and separating the oil. This greatly extended the bath time from days to several months, allowing higher concentration of oil for more efficient recovery, and recycling of both water and detergent. More continuous use of the bath system along with considerable cost savings through greatly reduced expensive detergent consumption as depicted in Fig. 8.3.

8.2.2.1 SE Technical Evaluation Review of Example 2 Parts Washer Case Study: Results after Installation of Ultrafiltration System (Fig. 8.3)

- Concentrated oil and grease (O&G) in water by a factor greater than 500.
- From 150 to 98,000 parts per million (ppm).

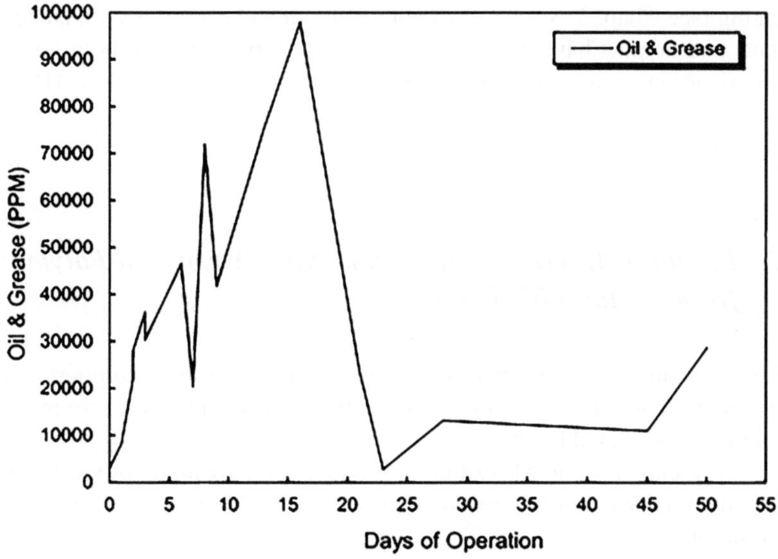

Fig. 8.3 Koch UF retention improvement

- Extended bath life from 1 week to over 3 months.
- Payback period was 6 months.
- Most of return on investment due to savings from designer detergent soap.
- O&G pumped to oil recycle for reuse elsewhere.
- Reduced water consumption.

8.2.3 Example 3: Process Improvement Plastic Film Process Change Energy Basis Review

A national manufacturer of plastic packaging materials and films decided to switch from a solvent-based to water-based plastic film manufacturing process. Their client, a national retailer, requires that all such changes meet their sustainability criterion. I term this criterion as "Btu as coin of the realm" and conducted a process modification review to assure this objective was met. The process change was indeed found to reduce its Btu footprint by over 75%. In addition it also provided considerable materials and environmental emissions avoidance and cost savings for the manufacturer.

This involved a detailed review of all components entering in as feed as well as being sent to a permitted vapor phase thermal oxidizer that prevented atmospheric solvent release. A before and after product and process energy content review was conducted including changes in energy consumed in the process were tabulated by spreadsheet. The spreadsheet itself, not shown here, presents a pictorial of the SE process and aids in identifying and creating species balances. The double-entry accounting (see Chap. 3, Sect. 3.4) of both material and energy only balances out if all inputs and outputs have been "accounted" for as was the case here. This could also be done in a process modeling program, e.g., Aspen Plus, HYSIS, or ChemStations, but a basic spreadsheet also affords a very simple and inexpensive method.

8.2.4 Example 4: ADM Process for Super Absorbent Polymers from Sustainable Crops

This involved our process bench translation of client chemistry to support manufacturing process scale-up. The following is an excerpt from adm.com. No proprietary information is presented herein.

This represents a prime SE example of a nontoxic, chemical product manufactured from a totally renewable resource grown the world over. Here are some company descriptions:

Copyright 2015 Archer Daniels Midland Company website

"ADM helps meet customers' and consumers' desires for sustainable products with our bio-based superabsorbent polymers for disposable personal hygiene applications. These products not only offer an environmentally sustainable alternative to petroleum-based materials, but also offer effective absorbency and moisture retention for baby diaper, adult incontinence, feminine hygiene, food packaging and other industrial applications.

ADM's patented, breakthrough technology using modified starch results in a high-quality, superabsorbent polymer that is hypoallergenic, non-toxic, safe and made from renewable resources. Our bio-based superabsorbents have physical properties similar to those of petroleum-based superabsorbents, and thus, do not require changes in the manufacturing process.

The ADM line of superabsorbents includes the newest generation of bio-based material, BioSAP™, which offers customers the highest level of total absorption, retention, and absorption under load available for bio-based materials."

ACS Chemical and Engineering (C& E) News, (www.ACS.ORG) page 26 SEPTEMBER 2, 2013

ADM PLANS ABSORBENT STARCH FOR SRI LANKA

Agribusiness giant Archer Daniels Midland has formed a joint venture with Sri Lanka based CIC Holdings and specialty chemical maker Chemanex to build a facility in Sri Lanka for starch-based superabsorbent polymers. The products, like competing petroleum-based polyacrylate polymers, are used in baby diapers and other hygiene products. The polymers will be made from locally sourced agricultural feedstocks. Dow Chemical and BASF are working on polyacrylate polymers made from bio-based acrylic acid.-MMB

8.2.5 Example 5: MVR Process for the Recovery of Aircraft Deicing Fluids

The MVR heat pump is an outstanding energy multiplier that has great efficacy, particularly where the bottom and overhead temperature do not differ greatly. In this case the MVR can be operated efficiently with a COP of between (5–10)/1.

Figure 7.2 in Chap. 7 shows a distillation column with an MVR used to provide the reboiler heat and also replace the condenser and boiler. Note that the glycol is removed by the distillation column and exiting vapors are compressed to create a vapor that will condense against the bottom of the column which is approximately 50% glycol. A portion of this condensed, purified water stream is returned to the column overhead as reflux, and the rest can be either sewered through a POTW system, but SE would dictate some better reuse, e.g., water for lawns or toilets. It should be noted that modern systems use propylene glycol, a nontoxic compound that actually is used as a food additive. The very large energy savings is offset by the purchase of a relatively expensive MVR compressor, but still payback periods can be a short as 3–5 years. Water is not free so if the recovered water can be used, a credit should be taken against the purchase of potable water. Another potential benefit is this recovered water is mineral-free distillate and may even have greater use and value for that reason.

8.2.6 Example 6: Glycol Concentrator

Excerpt from Inland Technologies Canada Inc., www.inlandgroup.ca

> *"A key component of Inland's glycol recycling process is the patented "Glycol Concentrator." This technologically-advanced, cost-effective system removes water from collected glycols, producing a concentrated glycol/water mixture with up to 60% concentration. Concentrated fluids are typically distilled and reused for several applications including as feedstock for production of aircraft de-icing fluid and as heat transfer and engine coolants. These self-contained units are based on mechanical vapour compression.*
>
> *The Glycol Concentrator outflow system has been fitted with a "scrubber" that reduces glycol concentrations to less than 1000 mg/L, meeting the most stringent environmental guidelines. In cases where we process material in two stages, the glycol in the distillate is reduced to 100 mg/L. If ultra-pure water is required, membrane treatment is used.*
>
> *The Glycol Concentrator is designed to operate with minimal maintenance and operational attendance. The computer controlled operating system has built-in safety systems and checks, further ensuring ease of operation. Since first being developed, the Glycol Concentrator has processed approximately 650,000,000 liters of effluent*

8.2.7 Example 7: Recovery of Zinc from Automobile Scrap Metals—Meretec Mittal

The following generalized description is also available on the web. No proprietary information is presented.

In brief, automobile metal parts such as steel chassis and sheet metal are often galvanized with zinc as a corrosion inhibitor. Zinc is an important metal and can be recovered and separated from the steel it is coating. This is a thermochemical process. Basically, shredded scrap metal is washed in hot caustic and the resultant zinc is recovered from solution by electro-winning. The cleaned steel scrap is valuable for immediate steel mill use once the zinc is removed and the recovered zinc metal of course is also quite valuable.

SE—Processes like this should always incorporate heat recovery and/or electric generation. Of course, location near an end user, e.g., steel mill, for direct incorporation of recovered products is the key to process economics. Both of the recovered metals greatly reduce the need for costly earth mining and represent a terrific SE methodology as is!

8.2.8 Example 8: Plastic Wood, Trex Inc.

This product is made from recycled plastic and wood. Per the manufacturer, it does not have sufficient structural strength for home building, but is used on decks and outdoor furniture as a cost-effective wood replacement. There are a number of products competing for recycled plastic. The manufacturer claims this product wears better and needs less maintenance than wood decking.

Also, waste products can be combined with wood wastes, e.g., chips and sawdust to make a nonstructural product suitable for some, if not all applications. Particularly outdoor applications in parks, homes, and nonstructural load-bearing applications, e.g., structurally supported porch decks.

8.2.9 Example 9: Paper and Pulp Production and Recycling from Confederation of European Paper Industries (cepi.org)

Several new technologies applicable to SE under review include:

1. Deep eutectic solvents (DES).
   (a) Dissolve any biomass into lignin, cellulose, and hemicellulose.
   (b) Recover cellulose from waste.
2. Flash condensing with steam.
   (a) Nearly waterless paper production.
3. Supercritical CO_2.
   (a) Dry pulp and paper without heat or steam.
   (b) Similar to coffee decaffeination.
4. 100% electric production.
   (a) Shift from fossil to electric fuel.
   (b) Store energy as hydrogen or pulp.

✈The gasifier is also available here for handling final, irreducible, and recalcitrant organic wastes. Paper industry waste quantities are large enough to support integrated power generation and chemical production.✈.

8.2.10 Example 10: Construction Debris and MSW Reuse at Military Installations

There is money to be made in P2—US Air Force ESOH-CAMP Example at Keesler Air Force Base, Biloxi MS of separating glass into various colors, newspaper, and cinder blocks into simulated coral reefs to support ocean fish breeding, metal into steel and aluminum, plastics by type, etc. Local recycling efforts are focused on MSW. This also helped keep oil and paint waste out of landfills. A novel reuse of construction materials in off-shore fish wildlife habitat was very successful in augmenting surrogate reefs (We made sure the resource manager received a reward for this good work).

8.2.11 Example 11: Yeast to Milk

According to *Smithsonian Magazine*, GMO (genetically modified organism) modification of yeast can create a product with milk-like properties. Whether this makes sense or not remains to be seen, but it does take a calculable effort to raise dairy cattle. Again a full analysis will vet this if and when it comes to pass. Applications claim to be aimed at Vegans for now. (https://www.smithsonianmag.com/smart-news/forget-fake-meathow-about-fake-milk-180954996/).

8.2.12 Example 12: Seafood Processing

This is a fascinating, totally renewable business that occasionally needs help from international governments to prevent overfishing, similar to preventing deforestation. Everything is used in this process and I mean everything. The Pacific Ocean, Alaskan plant I visited processes over two million pounds of fish per day when the fish are running. Essentially everything is productively used and even excess fish oil is used to provide clean burning electricity generation. This is a model SE industry when done properly. Even wastewater can be filtered to recover valuable fish protein before reusing or returning to the sea.

8.2.13 Example 13: Agricultural: SFP Fertilizer Nontoxic Fertilizer Enhancements

This was presented in Chap. 5 as an alternative manufacturing choice. I mention it again now for completeness. Here, solid, particulate-type fertilizers like the ones created in the "Corn to Polyols" design example at the end of this chapter, are coated with a nontoxic polymer to better control the movement of nutrients to the plant. This led to increases in production of 5–15% or so and also served to keep the fertilizer in the ground where needed by the crop instead of dispersing into the environment. This is not GMO. It is a great example of SE design for enhancing existing fertilizer utilization to minimize waste and maximize efficiency (We also helped this client with the USEPA PMN submittal process as described in chapter 5).

8.2.14 Example 8.14: Large-Scale Animal Farming—Too Large to Succeed!

Choose your poison—Whether you feel it is unethical to raise animals on a large scale, in confined space or not, it is clearly leading to severe challenges in the large-scale use of antibiotics and vaccines and the present business model is leading to outbreaks of incurable chicken virus with millions put to death—April 22, 2015, news item. On September 10, 2015, McDonalds Corp announced it was switching to egg products from free-ranging chickens in the wake of the catastrophic flock devastation in large caged chicken farming with antibiotics. Additional downstream effects include antibiotic resistance in humans as well. The matter of waste disposal and post-processing may actually benefit from large-scale practices if done properly.

The point is that consideration of these matters falls neatly under the category of sustainability engineering SE. No business wants to suffer periodic slaughter of five million chickens as occurred recently, with the additional threat of spread of an untreatable poultry disease. So, these types of considerations must be included in agricultural plant design regardless of the desire for free-range animal consumption. This is similar to the global warming concerns in that SE practices almost always lead to reduced CO_2 footprint, that is to say SE efforts nearly always have the unexpected positive consequence of reducing CO_2 footprint as well as increasing profits regardless of the driving force. Similarly, current mega poultry and pork farming is probably unsustainable and will need SE-designed alternatives to improve them, which almost certainly leans toward more humane practice. Consumers are already voting with their feet and pocket books by shopping at stores featuring non-GMO and antibiotic free agricultural products. Again, SE shows the way to a more productive, safer, *and* profitable future for business and consumer alike.

8.2.15 SE Agricultural Engineering Challenges and Opportunities

1. Water recycling—Easy to clean and reuse as irrigation and feed water.
2. Solid waste recycling—Composting and biodegradation produces fertilizer and energy.
3. Antibiotics—Design animal spacing, prohibit use as growth stimulant, no human impact.

8.2.16 Turning Farm Manure into Renewable Natural Gas

Excerpt from "Cows to Kilowatts: Black & Veatch Awarded EPC Project to Convert Dairy Farm Manure into Renewable Natural Gas":

Publish date: Wednesday May 18th, 2022 | OVERLAND PARK, Kansas

Engineering and construction leader to design-build system to curb dairy farm greenhouse gas emissions while simultaneously creating a new revenue stream based on sustainable fuel.

Although dairy is a dietary staple in many American households, it can be hard on the environment due to the amount of greenhouse gases (GHG) produced from dairy farms. To address this, Black & Veatch has partnered with Amp Americas to develop an engineering, procurement and construction project that turns cow manure into renewable natural gas (RNG).

According to the U.S. Environmental Protection Agency, methane is a powerful GHG, one that is 25 times more potent than carbon dioxide at trapping heat in the atmosphere, contributing directly to climate change. The International Energy Agency estimates that annual global methane emissions are around 570 million tonnes. Full-scale decarbonization will require innovative solutions to curb or reuse emissions current industrial processes — agriculture included.

Source: https://www.bv.com/news/black-veatch-awarded-epc-project, May 18, 2022

8.3 The Btu as the Coin of the Realm: A Key to SE

Take a moment to view some economic elements for SE. Any change that is made to either an existing operation or planned design must be fact-based and justified with follow-up metrics evaluation. To paraphrase Deming: "If you can't measure it, it doesn't exist" (see Chap. 9). And remember, the law of unintended consequences is a fickle partner. Sometimes we are pleasantly surprised, other times not. This is why the emphasis on quality circles and teams and overall quality management. This helps us "connect the dots" to avoid surprises.

I drive a Prius automobile. My 2005 model averages in excess of 45 mpg year round here in the Chicago region with wild climate variations of −10 to +100 F. This is nearly double the mpg of present federal mandated guidelines for US automobiles. But does it make sense sustainably? This vehicle requires an expensive NiMH battery that unlike the IC engine cannot be easily and cheaply rebuilt and will need to be replaced and recycled at some point. In addition, it has a high efficiency continuously variable transmission and computer control system. (Mine is 10 years old, with 130,000 miles.) Toyota has indicated they added value beyond the sales price of this machine, perhaps to encourage early adoption, perhaps to boost fuel average MPG ratings—if so, this may contradict or at best obfuscate a Btu-based SE evaluation.

Simply using a pure economic analysis is complicated by the large variability in the cost of disparate energy sources such as fracking produced oil and natural gas, foreign market manipulations, nuclear and fossil fuel–derived electricity as well as wind, hydro, solar, and photovoltaics produced electricity. But a true overall energy and material balance, with Btu-based evaluation has always been the stock and trade of the process design engineer and should always provide a point of departure for all designs.

The average consumer will vote with his/her pocketbook. But manufacturers like Toyota may choose to produce these vehicles and sell them at low to no profit, merely to meet or exceed fleet average mileage mandates. While this is not sustainable in the singular sense, it may well be an excellent SE model for the company overall and a good argument to minimize government intervention on the "how to do it" side of the regulatory equation. The point being all such process changes must pass the SE test. In addition, just as in the case of classic chemical production design, an appropriate life must be assigned to each alternative for proper evaluation. ✈ This is not and should never be a way to impose control over business, but rather to inform decisions that will, in the long run, be not only more sustainable, but profitable as well. Life cycle analysis (LCA) looks at "cradle to grave" disposition of project elements and can be a useful tool that in the SE design decision process.✈.✈Today's sub-$3/gallon gasoline is a losing proposition for anything but gasoline or possibly full electric unless the vehicles are comparably priced. I calculated about $6000 savings thus far, not quite the $10,000 I estimated at the point of purchase in 2005. But either way, nearly 2000 gallons of gasoline were saved.✈.

8.3.1 Plug-in Electric Cars

A similar SE economic evaluation must be made of all-electric cars. Keep in mind that an auto manufacturers' event horizon may only be 5 years at best. See Chap. 4 on efficiency. The economics of plug-ins must be reviewed carefully. Again, in 2016, $3 gasoline just does not cut it, even with steep government incentives to purchase more efficient vehicles. It is however, much easier to control emissions from a single, central power plant than from hundreds of thousands of tailpipes, a fact which also needs to be included in the SE evaluation. Electrics also displace the production of high octane fuels, saving additional value-added refinery processing costs. The 2016 glut on the market of natural gas was giving many technologies a run for their money. Some auto manufactures have dual fuel capabilities allowing LNG or gasoline. As stated before, these types of evaluations are amenable to spreadsheet what-if scenarios enabling the SE designer to easily adjust to market conditions. Note: High price of gas now makes electrics more attractive, but the final evaluation is bottom line, electrics get 3–4 times better mileage than gasoline.

8.3.2 Food Versus Agri-Chemicals Production

Most large agri-businesses such as Cargill and ADM make a host of useful agri-based chemical products. This of course presents the need to balance food versus chemical production, which they do quite well. But government agricultural subsidies can obfuscate the overall economics when applied to SE design. Not a daunting or impossible task, just a checklist item to put into the economics spreadsheet. Agriculture operates to a large extent at the whim of nature. Even the ancients recognized the need to maintain granaries to offset the bad years.

8.3.3 Food Versus Fuel

Corn to ethanol—represents a good example of government's attempt to reduce dependency on foreign oil. Although the Btu footprint does not yet justify this, much research has been spawned in developing ways to produce ethanol from cellulosic waste products and it is hoped this will succeed in the long run. Another important overarching planning element here includes the fact that the corn used for ethanol could certainly be pressed into food service if the need ever arose. This is a prime example of creating a bridge to long-term sustainability.

8.4 In Works: But "Not Quite Ready for Prime Time"

Actually, as I write this, some of these are starting to poke their head above the SE event horizon!

8.4.1 Algae to Oil

This is a holy grail of sustainable energy coupled with carbon sequestration, i.e., something for everybody, by converting CO_2 into useable hydrocarbons by the addition of energy, typically, but not necessarily sunlight. Very large tracts of water surface are needed in nature to support the summertime algae blooms that contaminate swimming areas in water with low O_2 content. Bioreactor companies are working on reducing this footprint through classical chemical and environmental process design engineering, now SE Design.

This SE technology might eventually supplant petroleum as a fuel, leaving crude oil for much higher value added end use such as plastics, or for that matter, it could do it all.

(Pacific Northwest National Laboratory) By Tuan C. Nguyen smithsonian.com December 31, 2013

Algae to crude oil—Researchers at the U.S. Department of Energy's Pacific Northwest National Laboratory have discovered a way to turn a small mixture of algae and water into a kind of crude oil in less than an hour.

There are a number of novel methodologies in the works worth watching, including:

JPP Note: According to Mongabay.com, Exxon Mobil and others have deemphasized algae biofuel, including scalability among other challenges. Wind and solar are running ahead but these are not necessarily suited to plant-scale operations. Bloomberg.com notes that Exxon Mobil is backing away for now.

This will leave national labs to pick up the slack to identify test, and develop robust productive algae strains.

8.4.2 Hydrogen and Ammonia Fuel Economy

Using NH_3 as a fuel has also been proposed. H_2 and methanol can be made from NH_3. The processes are also CO_2 free [1]. This process is integrated with syngas and H_2 production with CO_2 recycled to oil fields for use in enhanced oil recovery, removing the necessity of onsite CO_2 production. Not quite SE, at best an extender of natural gas and SE bridge, which is highlighted in Appendix C.

8.4.3 Using Local Green Energy and Ammonia to Power Gas Turbine Generators

Here, ammonia is used as a transportation fuel as well as an off-peak large-scale energy storage medium. Proposed convert from oil- to ammonia-powered gas turbine generated electricity (Chap. 7, Sect. 7.4). (Also under development is artificial photosynthesis.)

8.5 Open-Ended Questions that Need to Be Answered by Sustainability Engineering

Does human activity cause global warming? Can we antibiotic our way toward ever-increasing sized farms? Turns out SE might have some answers:

1. It does not really make sense to pump carbon into the atmosphere, the stuff is too precious to waste that way. Eventually it will get absorbed by plants through photosynthesis, but not fast enough neither for the business folks I know nor to avert global warming. Gasification of MSW to create CO and H_2, for instance, especially with integrated power and chemical production, of nearly any organic, minimizes CO_2 footprint of generation and refining and saves money. Hydrogen fuel cells will also figure into motive power and electric generation. SE will save money, just as pollution prevention did in the 1980 and 1990s—It does, however, take good science along with smart engineering.
2. There is no question that antibiotic-resistant strains of bacteria are on the rise, no matter what your politics, so we are going to need other means of raising healthy farm animals without antibiotics for long-term agricultural sustainability. Farm wastewater treatment is also amenable to SE and will save additional money on all of this if engineered properly.
3. More development, both technical and regulatory is needed to support the flow of electricity both from the grid and back into the grid. One cannot simply connect rooftop solar panels or industrial peak generators to the grid as though it were an electric socket. Sophisticated controls are needed to prevent damage as well as allow for billing/crediting to take place.
4. Regulations for the combustion of organic waste materials will need to be modified to include the production of power, while still protecting the environment.

Really folks, we will leave the politics to others, but there are some interesting SE-type questions that can be best answered by good science and smart engineering. SE jumps around the politics and presents a best practice way forward. In the process we will find more sustainable, cheaper solutions to the following. I am sure you can think of many others. Here are a few.

8.5.1 Alternative Fuels

Regarding plug-in electric cars, for instance, from an environmental perspective, a nationwide massive increase in electricity generation capability to supply automobiles would be needed. As I wrote this in 2016, Tesla was addressing part of this in planned battery production and charging stations in Nevada and California, now moved to Texas. But electricity generation and power line transmission, both largely inefficient would also need to increase. A positive outcome, however, is the relative ease of capturing point source power plant emissions as opposed to millions of tailpipe emissions including carbon sequestration. In addition, from an otherwise nonintegrated central electric power generation CO_2 could then become feed for a solar-based algal oil production if collocated in SE design style adjacent to power production plants. Presently, coal might also be suited for this purpose, but only if the by-product mining and processing wastes can be effectively captured and harnessed for reuse as described elsewhere (gasification might work here as well as a clean alternative to coal combustion). But this presupposes an acceptable method of mining or returning mined lands into useful or native habitable environments. Shale oil and gas-fired ones might be more amenable to environmental consideration and these are strong considerations that can be dealt with if the quality circle is drawn around the mine as well.

If all this seems daunting, it really is the domain and responsibility of the process design engineering team that has drawn a large enough quality circle to include these elements. For instance, as refinery output diminishes in an all-electric vehicle world, so too does asphalt production (see Chap. 11, Sect. 11.5.2). The specification for production of asphalt for road use, a refinery by-product, is actually strictly controlled by Department of Transportation (DOT) standards, federal, state, and local. Those standards are largely responsible for the relative safety of using these materials on thousands of miles of roads throughout the USA. The CPI has always been faced with fuel choices, e.g., natural gas versus oil-fired boilers. SE design methodology will also fine-tune these choices.

8.5.2 Algae to Oil: Combined Solar and Biotechnology

Algae can be grown by an anaerobic process akin to photosynthesis of CO_2 and light, and a by-product of this is oil. The holy grail of eventual application will revolve around the development of either batch or continuous processes that require 3–4 magnitudes less area as do the naturally occurring algal growths which presently poison drinking water supplies during summer months with stagnant water conditions. This will be a combined scientific and chemical engineering challenge. My senior design class took this on as a challenge project one year and ran up against the need to flood a major portion of a Midwestern state to meet algae to oil production needs … yup, seems doable, but needs more engineering!

Here is an example of ongoing algae-to-oil R&D:

Excerpt - RICHLAND, Wash. Pacific Northwest National Lab (PNNL)

Engineers have created a continuous chemical process that produces useful crude oil minutes after they pour in harvested algae — a verdant green paste with the consistency of pea soup.

In the PNNL process, slurry of wet algae is pumped into the front end of a chemical reactor. Once the system is up and running, out comes crude oil in less than an hour, along with water and a byproduct stream of material containing phosphorus that can be recycled to grow more algae.

✈I always take reports like this with a grain of salt, but if this is just partly true it represents a major advance forward for SE.✈.

8.5.3 *Solar Energy Electric Capture: Where Are we Now and where Are we Headed?*

This requires large footprint/kW-h and it may be better for CPI to use optically concentrated solar heating on rooftops, or large open fields typical of the unfertile southwest USA to generate steam for process power and heating or low-level "district" heating. The national grid can be used as a sort of storage battery or topping cycle to allow large-scale electric generation to be economical (Chap. 8, Sect. 8.7). A favorite class problem is to balance pumping water to elevated storage during off-peak with on-peak letdown generation. Depending upon local utility and geographical situation, this can be economical.

In the end, large solar electric generation will offset other fossil fuel usage during the day for use at night, a huge SE savings factor—the battery effect!

8.6 Recycling: A Key Component of Sustainability—Common Success Example Already in Place

Used automotive oils, dry cleaning fluids, batteries, and restaurant deep fry fat are all examples that are mentioned here only to demonstrate what can be done [2]. The business that has grown around this is enormous, successful, and the model can be transferred to the CPI. (I know, I worked in it for 2 years prior to returning to graduate school!) It may even turn out to be more economical to use the automobile crankcase oil as feed for gasification-based chemical production instead of reentering the refinery. Wood pallets, metals, plastics, and glass are a few examples of the move toward renewable resources. My USAF experience teaches that much more can be done, particularly in recycling MSW.

8.6.1 *Food Recycling: The Forgotten SE Element*

All food should be recycled in some manner, whether as food or fuel, do not throw it away!

- Food pantry whenever possible of course.
- Garbage to energy only.
- Frying oils to oil recyclers.
- Biodiesel production.

8.7 CO_2 as a Feedstock: Competing with Algae, or Showing the Way?

This represents a novel, SE-renewable augmentation of CO, H_2, and otherwise waste materials. Some of the company's processes use CO_2 as a feedstock over a proprietary biocatalyst. Thanks to Dr. Michael Schultz of LanzaTech for the following.

8.7.1 LanzaTech Process Description (by Dr. Michael Schultz, LanzaTech)

LanzaTech's basic platform uses the carbon in CO-containing waste gases, including steel plant stack gases, as feed for a broad range of fermentations products such as 2-carbon to 4-carbon alcohols and acids and/or other renewable chemicals and fuels. A proprietary reactor system ensures efficient distribution and conversion of the gas by the fermentation media. Various industrial waste gases, such as those from a steel mill basic oxygen furnace (BOF) as well as other primary ferroalloy process stacks can be exploited. Typically, such a gas stream is 45–50% CO, with N_2, CO_2, inerts, and dust.

As shown below in Fig. 8.4, the waste gas is passed through gas handling to remove particulate matter and oxygen. The gas is injected into the fermentation reactor, which is filled with aqueous media and microbes, where it is consumed and converted into ethanol or other metabolite products in a single pass. The product-rich broth is continuously extracted from the bioreactor and passed through to a distillation unit. Water and media components are treated and recycled to fermentation. Products are transferred from the distillation to product tankage, where they can be loaded onto rail cars or trucks for delivery, or alternatively piped to a nearby facility for the conversion to higher value chemicals.

8.8 Integrated Power and Production in the Chemical and Related Industries: Secret Santa of the CPI

Centralized, offsite utility companies' electric power production can lose 60–70% in terms of conversion efficiency and transmission line loss from fuel to plug. Chemical manufacturing employing large amount of heating and cooling can be quite inefficient as well. Modern manufacturers have learned how to accommodate these facts, indeed to absorb them into the bottom line total cost of doing business. ESOH compliance has also been largely accomplished as well. But is there room for improvement?

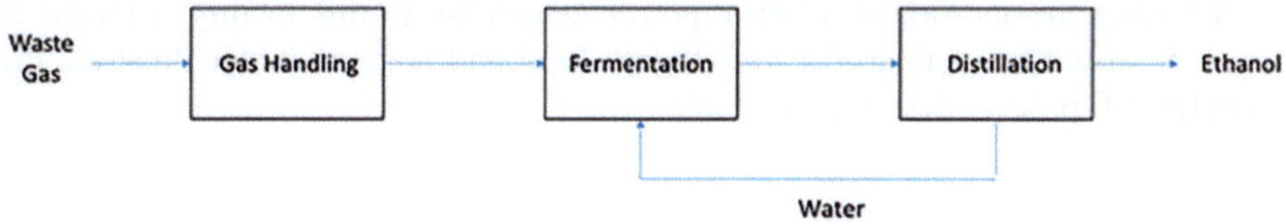

Fig. 8.4 Block diagram and description of LanzaTech process. (Courtesy Mike Schultz and LanzaTech, 2015)

Common CPI manufacturing methods produce an irreducible level of waste energy and material, the disposal of which has also been accounted for in the bottom line. Nonhazardous wastes of all types, organics in particular, are present in foods, pharmaceuticals, petroleum, and chemical production in large quantities. As we have seen, gasification of organic waste material can be used to produce H_2 and CO and heat source/sink. But incorporation of plant electric power generation and sales to offsite users, can open new efficiency pathways.

There is little to no reason for any organic material, be it chemical or municipal waste, to find its way into landfills. Either gasification or pyrolysis can be employed to convert these materials into basic building blocks of the CPI, e.g., H_2, CO, and various other chemical intermediates. Some enlightened regulatory development might be required to deconflict/decouple these from incineration.

8.9 So Putting It All Together Stepwise

Here are a few examples demonstrating the progression from basic SE material and energy-saving chemical process manufacturing, to fully integrated power and production.

8.9.1 Step 1: Combine and Interconnect Disparate Chemical Processes—Initiate SE Design with Material and Energy Balance

- 1. Minimize heating and cooling losses
- 2. Reduce Raw Material Use
- 3. Natural gas gasification- (reforming-) based process.

As an example, Fig. 8.5 depicts a combined chemical and fuels production facility. Gasification of natural gas is featured. Heat and materials utilization is greatly enhanced, leaving high value-added merchant products (Gruber, George, "Reforming Natural Gas to Chemicals and Clean Fuels," Personal Corp Communication).

A further SE addition of electric power generation to this facility depicted in Fig. 8.5, would bring us to the next higher level and helps to better illustrate the complete SE process design procedure.

8.9.2 *Step 2: Integrated Electricity Generation with Chemical Processing—Completing the SE Circle*

1. Provide overall excess organic fuel feed, e.g., natural gas, to chemical production train.
2. Generate sufficient plant electricity.
3. Generate additional grid electricity.

(a) Used for income as well as surge capacity (load balancing).

4. Generate additional natural gas for local use.

(a) Used for income as well as load balancing.

5. Integrate water usage wherever possible.

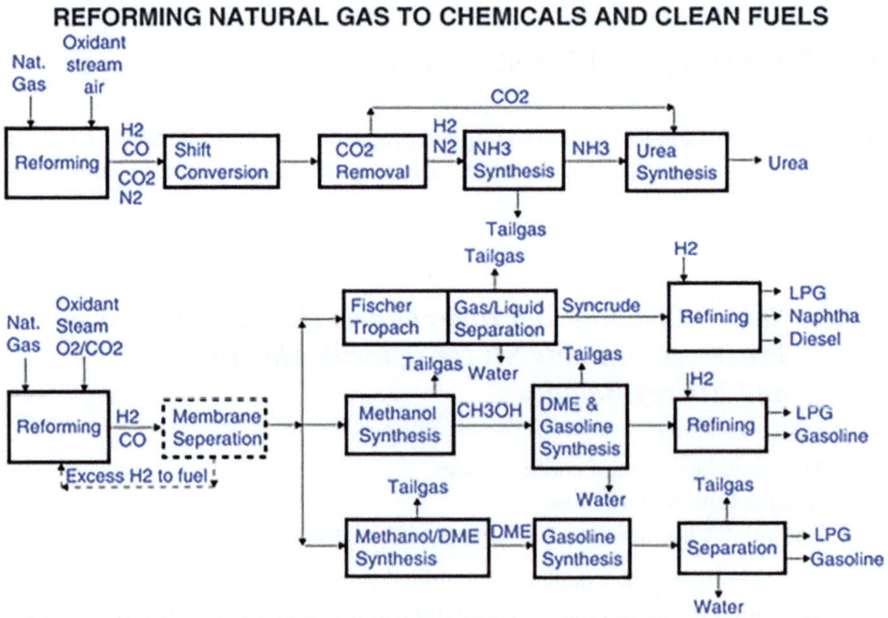

Fig. 8.5 Combined chemical and fuels production facility: SE minus the power integration. Courtesy of George Gruber, Black, and Veach

As previously noted, electric grid interfacing requires controls for voltage, frequency, and reactive power damping. Power factor, or quality, is also an issue. Not insurmountable, just a design checklist item.

8.10 Fully Integrated Power and Chemical Manufacturing Examples

Gasification is the one-stop chemical conversion and feedstock generating process technology and has been used for around 100 years to produce various fuels and other chemical products. We will take a look now at a few applications to convert natural gas into a veritable cornucopia of manufactured chemical goods. This technology both generates and consumes heat, but when combined with plant power generation can provide economies of scale all its own. In addition, low-quality heat not typically useful to make power or steam, can be captured in classic, chemical process economizer mode. This can be optimized through judicious selection and pairing of power, gasification, and suitable chemical manufacturing modes.

This is what the integrated power and chemical processing plants is all about. I used several of these examples as design problems in my senior design class that were literally "ripped from the headlines" that we will look at shortly. They were a great way to help get the students inculcated to the SE design approach! Key to success was having ten of my local area engineering design colleagues, who helped create the spring semester problems as well as mentored the students.

A note of caution, however, planning onsite grid interconnection to accommodate power generation requires close pre-planning with electric utility. Cannot just simply flow electricity backwards! Our electrical engineering colleagues will create that controlled interface.

8.10.1 Example 15: A Fully Integrated Power and Shale Gas Chemical Production Complex (Shale Gas to Chemicals)

One thing they have got quite a lot of in Williston ND is natural gas. By adding more natural gas than needed for chemical production as shown in Fig. 8.6, one can also produce power for the facility as well as surrounding community. The external connections serve also as a surge capacity topping cycle which can allow for swings in costing on either side of the power or chemical production efficiency curve. My Chicago engineering colleagues and I used this example in teaching my spring semester senior chemical engineering design course (Appendix B). We also added the Midrex process (Chap. 3, Sect. 3.7) for direct iron ore reduction (DRE) to aid in a more complete material and energy balance. A more complete description is found in Appendix C. This served as an excellent training ground to demonstrate SE design principles to our young students. I am indebted for the assistance of my Chicago area colleagues, listed in Appendix C.

In Fig. 8.6, natural gas and iron ore are the raw materials. By locating near the wellhead, say in Williston ND, the plant is situated on rail and highway for easy movement of merchant products. Also, by locating adjacent to a steel processing mini-mill, hot reduced iron can be fed directly saving quench energy costs.

8.10.2 *Example 16: A Fully Integrated Power and Corn to Polyols Production Complex*

In another great SE design example from our class, we focused on "corn to polyols." Here, in Fig. 8.7, corn on the cob is the feed, from which all other products are made. Traditional wet milling provides common merchant products. Excess corn feed provides sufficient production of dextrose to sorbitol to polyols production. I believe we located this one near Decatur, Illinois.

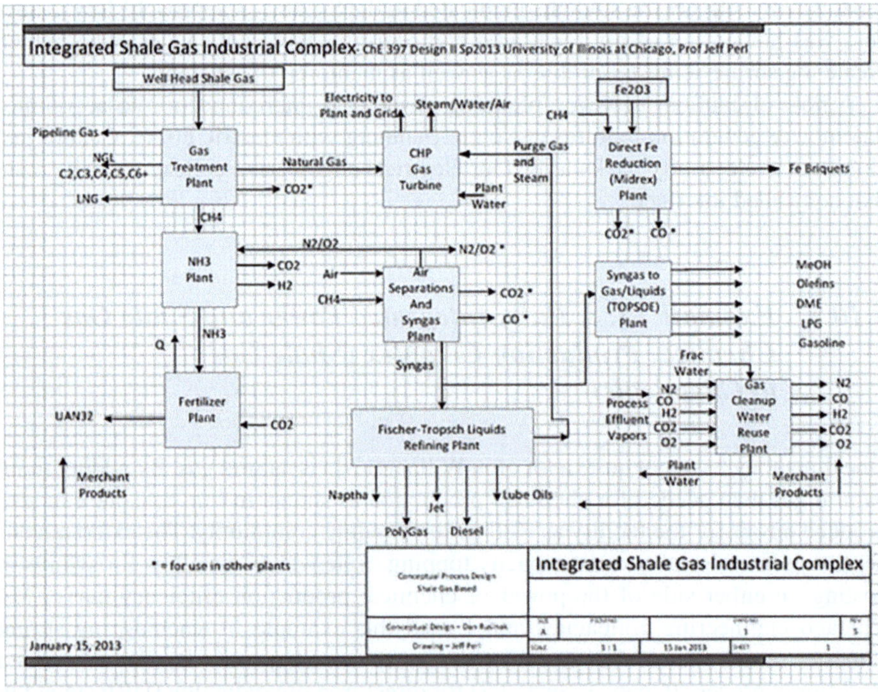

Fig. 8.6 Integrated power and shale gas chemical production complex

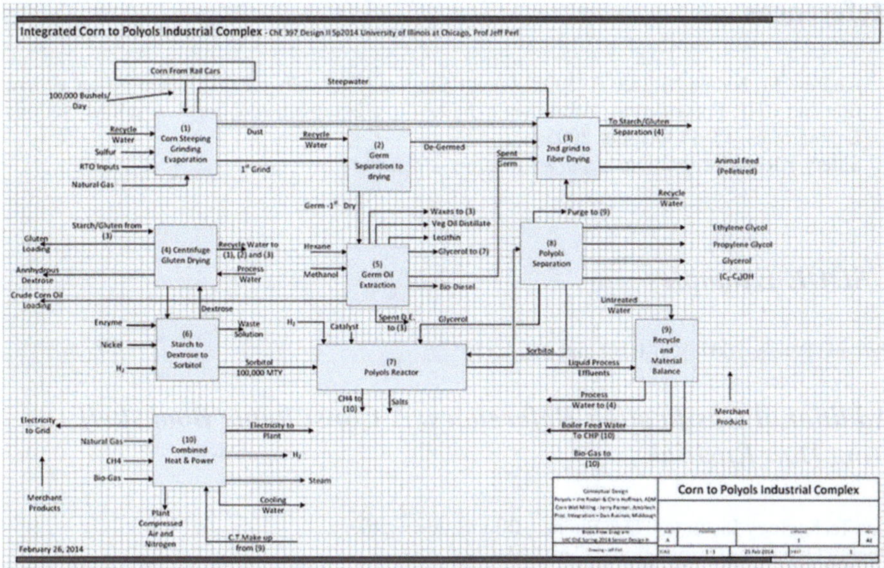

Fig. 8.7 Integrated power and corn to polyols production complex

While the argument of food versus fuel is pertinent, this represents a nearly perfect SE design in that the raw material, corn is truly renewable. The others are also great SE candidates as they all greatly extend the M&E equation. Of course, with gasification at the nexus of energy and chemical production, coal and oil are also candidates, albeit nonrenewable resources. Other possible methodologies include biomass reactors, algae to oil, and other renewables.

8.11 Summary: The SE Application Bottom Line

The key again is to colocate disparate processes to maximize material and energy utilization, while minimizing the overall SE footprint. Integrating power may not always be necessary, but it sure makes the most sense. It is also an ideal use of an excess supply of natural gas. The 40% HRSG power generation efficiency is increased considerably by SE design integration of heating and cooling of both low- and high-quality heat, realistic only in chemical production. Transmission line losses all but disappear, and this is one big electric power topping cycle that uses the grid as a storage battery (Alain Semet, Personal Correspondence, 2014).

References

1. Andrew M, Nicola M, Luc P (2005) Direct ammonia fuel cells for distributed power generation and CHP. Presented at "Ammonia—the key to a hydrogen economy". Argonne National Laboratories, 13–14 October 2005
2. Safety-Kleen Corporation. http://www.safety-kleen.com/

USDOE Examples

Perl and Peters, 2007, [Ch 1:10] and [Ch 1:11]

Additional Gasification References: A Consultants Bookshelf from Dan Rusinak

Gas Technology Institute—Des Plaines, Illinois, www.gti.org
Energy Dynamics Corporation—Carlsbad, California
Gas Technology Institute—Des Plaines, Illinois
Stamicarbon: Ammonia Processes
Higman et al. Gasification and Ammonia
de Klerk: Fischer-Tropsch Refining
Twigg—Catalysts

Industrial References

Gruber, Black and Veach, SGO3 paper, Natural gas conversion alternatives, Black & Veatch, 06 March 2012
Department of Energy-www.DOE.GOV

Additional Resources

Dickson O. High efficiency modular chemical processes (HEMCP), modular process intensification framework for R&D targets. DOE advanced manufacturing Office, 27 September 2014. energy.gov
Interconnection of large-scale photovoltaic systems with the electrical grid: Potential issues, Feb 2013 IEEE international conference on industrial technology, Mirhosseini, Mitra and Agelidis, Vassilios

Chapter 9
Total Quality Management and Sustainability Engineering

9.1 TQM and Similar Quality Management Methodologies

I have had a long-time interest in total quality management (TQM) and similar quality programs, and the following is from the last 34 years of experience. In 1994, I co-chaired, along with the American Society for Quality [1], a symposium on pollution prevention and quality management [2]. Managers from Burlington Northern Santa Fe Railroad, UOP, Motorola, United States Air Force Center for Environmental Excellence, and the USEPA all presented the manner in which their respective organizations used quality management in the pursuit of pollution prevention and waste minimization. ASQ provided a generic quality management training program.

I am including this section as I believe sustainable engineering principles involve more than just the client/contractor corporate engineering team, but the entire affected community. Also, any design team, SE or classic, that does not understand TQM, or its various cognates, is doomed to failure or at best being cast out to wander forever in the land of the suboptimal!

9.2 Quality Management and Pollution Prevention

Quality is a feature central to the success of every modern operation. In addition to Manufacturing, Pollution Prevention and Waste Minimization must be a part of every quality program. To make generic quality programs work requires specific technical inputs. An environmental model along with training specific to environmental awareness, and pollution prevention rounds out the requisite program elements. The symposium presenters, drawn from industry and government, will develop and present these system concepts. Although practical pollution prevention examples are important for illustration and will be included—the major emphasis

J. P. Perl, *Sustainability Engineering for Enhanced Process Design and Manufacturing Profitability*, https://doi.org/10.1007/978-3-031-52363-2_9

will be on the development of a fully integrated system of Pollution Prevention and Waste Minimization through Quality Management and Training Programs.

We are all greatly affected by this ongoing process. The Presentations and Panel Discussions with key Managers from the following organizations are designed to interactively illuminate these concepts in the great tradition of continuous improvement. [2]

We do not live in a centrally planned economy thankfully, as history has proven this to be suboptimal at best. The successful modern company operation is (or should be!) based on one form or another of TQM. This method recognizes the equality and importance of input from all direct activity participants. This includes the affected community as well as others having involvement. This requires drawing a quality circle to define those involved in the affected community, but also remains open to including those outside to join as well. The quality circle maximizes participant input within a structure but does not promote anarchy. After all, at some point a decision has to be made, plans implemented, observations and measurements performed to test the plan outcomes in comparison with objectives.

The use of quality management models is now ubiquitous. They are the ideal starting point for any discussion on sustainability engineering as they are by design ready to accept as input new ways of process improvement, including team-based methods to reject out-of-hand options that do not pass SE muster.

Here, SE-based economic model objective functions can be developed as they were pre-SE for evaluating options, as well as operating new, existing, and planned facilities.

9.3 A Generic, Basic TQM Problem Resolution Checklist

1. Define the problem.
2. State objectives.
3. Develop plan to meet objectives.
4. Implement plan.
5. Test and measure results.
6. Loop back for continuous improvement as needed.

These steps are not performed in a vacuum. The team must be comprised of interested parties, or stakeholders, with a quality circle drawn around the affected community.

This approach is amenable to any process where the desired outcome is an improvement over the existing one. While a lead or "champion" is typically appointed from among and by the team members, similar to a jury foreperson, TQM recognizes the importance of receiving input from all involved parties. Adherence to testable facts is required. As Deming said, "If it cannot be measured, it does not exist."

Some are small problems and pre-implementation is simple. Some are much broader in scope requiring additional input to assure against unintended consequences, particularly those arising from unknown unknowns.

9.4 Rumsfeld's Rules: Known Unknowns Versus Unknown Unknowns

Regardless of political philosophies, Don Rumsfeld used these rules in all the major entities he led [3]. The basic one revolves around his observations that there are *things we know* need to be determined in order to operate and solve problems, and *things we do not even know* that ultimately will affect the program. In other words, you cannot know what you do not know … unless you already know it all!

Being open to this reality has always been important, but more so with SE which is far less forgiving of leaving out game changing considerations. Quality management principles can provide real-time incorporation of new information by encouraging continuous improvement throughout the entity, whatever it is.

9.4.1 Known Unknowns in the Process World

Such as temperature and pressure of a stream not yet measured, but needed for an energy balance. Among others, we know we need to know:

1. Physical properties of all chemicals Involved.
2. Quantities of chemicals used.
3. Production totals.
4. Energy consumption.
5. Cost of maintenance and operations.
6. ESOH and operations permissibility.
7. Availability of staff.
8. Availability of feedstock and energy.

Those are but a few of the "usual suspects" the SE design engineer will need but not necessarily have fully quantified at the outset of the design effort.

New, SE additions to the list will include:

9. Broad-scale effect of manufacturing beyond fence line, e.g., unregulated items.
    (a) Overfishing.
    (b) Deforestation.
    (c) Overdrilling for oil and water.
10. Renewability of feedstock, e.g., petroleum, natural gas, wood, minerals, and metals.

11. Renewability of energy, e.g., petroleum, natural gas, nuclear, and solar.
12. Ability to integrate process and power.
13. Ability to integrate disparate processes to maximize heat and material conservation.

9.4.2 Unknown Unknowns

"You cannot know what you do not know." So, the objective of the continued advancement of chemical engineering or any discipline or profession is to achieve a more complete science-based understanding of that which you are studying. To beat down and minimize any unknowns to below a level of detection so as to have a minimum external effect on the process is the goal here. It is sometimes useful to think of the analogy with systematic, physics-based variations versus random, statistical ones. So again, the goal is to eventually discover and reduce all variations arising from seemingly random unknowns. Once the fact-based reason becomes known, it becomes part of the science-based model describing the system. In a perfect manufacturing world, all deviations from the desired outcome would be random, uncorrelated, and too small to be of importance. In the process world, upper and lower limits are set and a process is said to be "in control" only if deviations lie between those two limits [4].

Successful manufacturers already know how to incorporate continuous improvement so as to increase profits, reduce accidents, and prevent offsite emissions. But this is too often an after the fact activity. And by the way, there is no better way to discover the unknown unknowns than to steadily gather all the known unknown information and lay it out on the table and see what does not fit. You would be surprised at what will jump out after concentrated quality team study effort, sort of a form of brainstorming only in this case, with factual data as opposed to strictly emotional inputs.

9.5 Perl's Observations on Quality

1. There are two kinds of activities, optimal and suboptimal … not very deep!
2. There are two kinds of people, planners and schemers … now that is deep!
3. Planning is a quality activity involving the entire affected community within the quality circle.
4. Scheming has an exclusive non-quality limited circle composed at best of those who believe they know better, and at worst by thieves.
5. All non-quality-based systems are by definition suboptimal … or very lucky!

9.6 Going Forward with Quality Management Plans

One interim requirement is to generate a list of prescribed activities set out in the plan developed from the TQM checklist. This so-called "What needs to be done" list is laid out by the quality team, but the "How to do it" is left to the implementing team. This does not preclude the inclusion of a "Past How to Do It" information library, nor does it preclude such bedrocks as existing codes and standards that include specific "How to do it" legal requirements. Desired outcomes are clearly laid out in the plan, and are measured after plan implementation.

9.7 OSHA Process Safety Management (PSM): An Original CPI Quality Management Program

PSM is a fundamental regulation from OSHA [5]. This was in response to several explosions and catastrophic toxic chemical releases that resulted from preventable incidents. These incidents led to death and injury and the popular slogan: "safety is no accident." The old method of designing safety into a process relied on recognizing the obvious, or "low-hanging fruit" and if available, any accident scenario history. Low-hanging fruit, e.g., the need for tank pressure relief valves, was realized from the earliest days of steam power to prevent explosions. Later though, the complexity of chemical processes increased and it became necessary to perform fault tree—"what if" type of process analysis.

PSM requires that a process hazard assessment (PHA) be conducted to assure the process is doing what it was designed to do and has no pathways or operational modes that could lead to unwanted accidents or exposures. To do this, at a minimum, one must have block flow diagrams (BFD), outlining the overall process intent, process flow diagrams (PFD), showing stream flows with material and energy disposition at each point along with conceptual process control. Finally, the as-built piping and instrumentation diagram (P&ID) is absolutely required showing all pipes, valves, equipment, etc. depicting the entire process. Common PHA discoveries include misconnected lines, valves that should not be present, lack of pressure relief, incorrect safety procedures, and incompatible location of materials, to name a few.

Procedures clearly laid out for both operations and maintenance are checked against all the above diagrams and compared with fault analysis what-if scenarios to prevent the possibility of unintended negative consequences.

As a simple example, and without reference to any drawings, what happens if valve V-108 is left open while Reactor R-560 is operating? Each possible fault has to be accounted for. A team is assembled to perform this process hazard assessment (PHA). This is written up and all safety measures necessary are added on to the process, such as pressure relief, overflow or underflow alarm, etc. In addition, a method is found to manage any process changes for effect on safety and operability.

OSHA refers to this as management of change, or MOC, which is a requirement of PSM.

9.8 What Dr. Deming Taught the Japanese on Total Quality Management

In the forward to the 1990 book, "*Dr. Deming—The Man Who Taught the Japanese About Quality*" by Rafael Aguayo [6], Deming identified a lack of focus on quality as a prime cause of reduced US business. Deming went to Japan during the post-World War II US occupation, where he taught their top management and engineers his system of TQM. TQM lead to a huge surge in quality of Japanese products, e.g., automobiles, televisions and electronics. This eventually fed back to the USA, but not in time to save the television and electronics industries. Quality worldwide has since increased, as a result of Deming's systems approach to management that removed internal competition and replaced it with a team approach to benefit the entity.

9.9 In a Nutshell, Deming's Approach Regarding Quality Management Based Manufacturing

- Create a flow diagram outlining the system of the industry of interest.
- Estimate the market for the specific planned product or service of interest.
- Design that product or service.
- Mockup, test, and modify designs as needed through continuous improvement.
- In a well-organized system all the components work together to support each other.
- In a system that is well led and managed, everybody wins.
- Competition within the team will destroy the system, increase cost, lower quality and market decline.
- Top-down ranking with awards at top and punishment at bottom always fail.

The Japanese learned these lessons very well and within 20 years were creating high-quality manufactured goods including cars and electronics, two of their market-targeted industries. This was so successful that the US domestic consumer electronic industry including televisions, radios, phonographs, and recording machines to name but a few, all but vanished, replaced by higher quality, often less expensive imports. Free market forces eventually prevailed, forcing quality improvements by US automobile manufacturers that are still struggling to catchup with continuous improvement. All in all, this was in the long run a good thing for domestic as well as global consumers, and serves as ample reminder that all non-quality-based systems are suboptimal at best.

9.10 Quality Management Case Study: United States Air Force

Quality Management in the United States Air Force Installation Restoration Program—"Fixing Dysfunctional Remediation Teams," Jeffery P. Perl, AIChE Spring Conference, Atlanta, 2000 [7].

The generic quality management model below was based on experience working on HQ Air Force Center for Environmental Excellence, AFCEE environmental cleanup programs, aka remedial action.

As a bit of background, by act of congress, Defense Environmental Restoration Act (DERA) all military installations are required to meet USEPA cleanup standards just like the industry is. Two types of installations exist, namely those remaining open subject to the congressional mandated Installation Restoration Program (IRP) and those to be closed under the base realignment and closure program (BRAC). Both need to be clean, with particular emphasis on those undergoing closure and return or transfer to civilian use. Air Force Base and Installation cleanups commonly include soil contaminated from leaking underground storage tanks, mostly Jet-A fuel. In addition, chlorinated hydrocarbons, as well as heavy metals, were used in engine rebuilding. These subsurface contamination profiles are quite similar to most US airports in many respects. The Air Force established formats for productive relations between regulators, community, and the base.

My quality management methodology, depicted in Fig. 9.1, is translatable to any problem-solving requirement, in any area of endeavor. This one is related to the

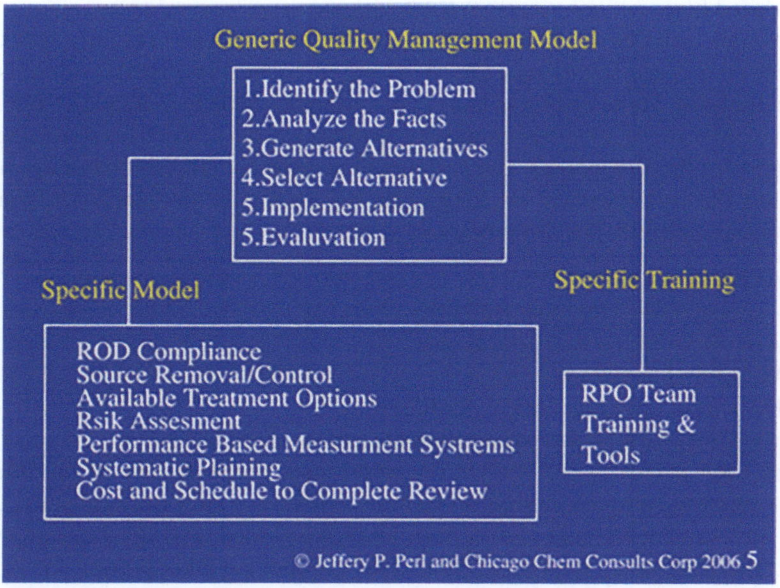

Fig. 9.1 A version of a quality model approach I developed about 22 years ago.

numerous BRAC and IRP cleanup programs within the Air Force. The elements are keyed to the figures that follow.

Element 1—The engine for this is the generic TQM, or quality management model, which is a fairly straightforward method for solving any problem. The keys here are reliance on fact-based decision-making (Chap. 1, Sect. 1.1), establishment of a quality team that includes the "affected community" and resources and commitment from management requisite to implement and measurement of the outcome of the team recommendations.

Element 2—In order to apply this method, a specific model must be built that covers all aspects of the process to which improvement is aimed. This diagram depicts some, but not all of these components. Here, a legal record of decision (ROD) issued by EPA is basically a prescriptive agreement entered into by all parties to outline, essentially in contract form, what the cleanup will entail. As can be seen, many elements are necessary and without organization into roles and responsibilities, success is not likely.

Element 3—Finally, the glue that holds this together and is requisite to success is training. Here, training consists of two parts, namely generic quality team building and mission-specific technical ones. Team building and rules of engagement are essential to overall management. Specific technical training is provided to fill in knowledge and experience gaps as well as assuring contemporary solutions.

With this generic quality process model, all team members can participate and contribute to the creation of an optimal overall solution.

A key program element, without which leads to dysfunction, requires assembling the quality team. I have seen problems whenever the quality circle is incorrectly drawn. Unfortunately, this is often done purposely. Figs. 9.2a–d are reproduced here.

Yes, this actually does happen if not controlled. Humans naturally will create groups, or cliques. Sometimes this is based on friendships, shared interests, and compatibilities. Sometimes it is based on a desire to come to a unique solution that does not involve the affected community, and this is what I refer to as scheming. The formation of the quality team can and should account for all the good things people do when they get together—so long as all have a seat at the quality table and all opinions are received for resolution inclusion.

Abbreviations

LRA—Local Redevelopment Authority

RAB—Restoration Advisory Board

PA—AF Public Affairs

BEC—Base Environmental Coordinator

Team Chief—HQ USAF designated project manager

The LRA is part of the civilian community and many former AF installations were successfully transferred at no cost to civilian use after being cleaned up, so LRA involvement was a key to success. The RAB helped make final cleanup selections. The BEC along with team chief watched over the process from the AF side and the PA Office made sure the plans were made available to the public, often at

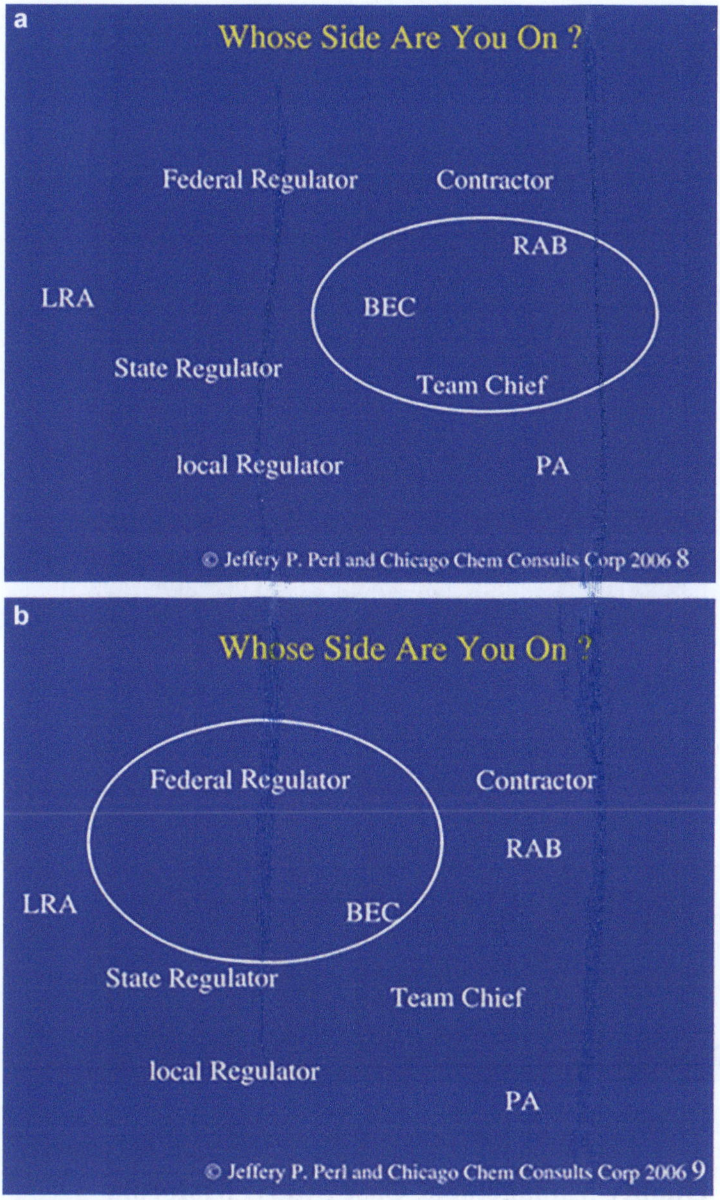

Fig. 9.2 (**a**–**d**) Drawing the quality circle—whose side are you on?

the local public library or online. You can see that it can be quite involved to obtain input from all participants, but in the long run, it leads to much better outcomes. Austin Texas' airport is a good example, as are large parts of O'Hare airport just to name a few.

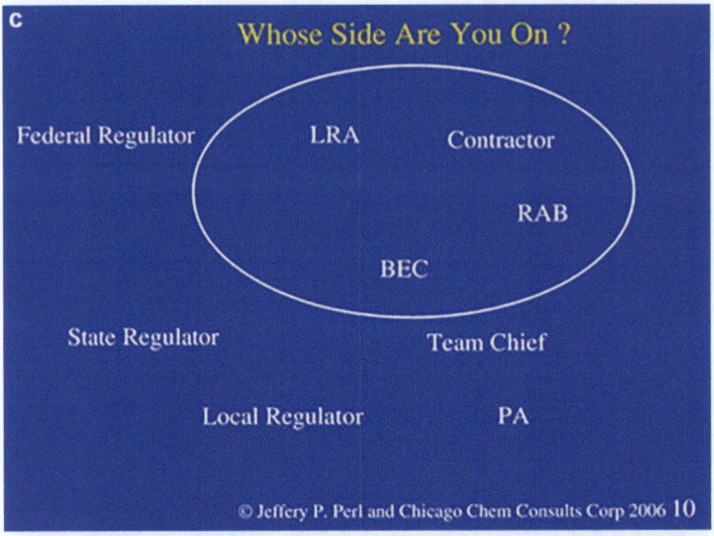

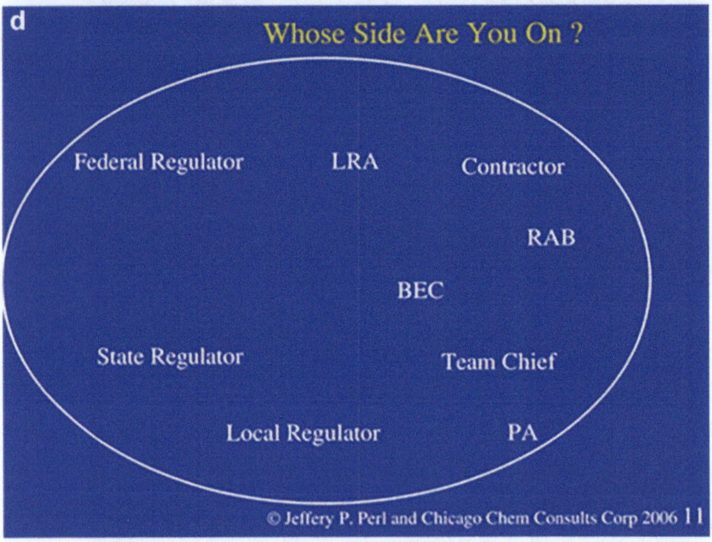

Fig. 9.2 (continued)

9.11 Quality Management Case Study: Motorola Electronics Production

Another example, related to sustainability is one I created some time ago. While working with Motorola HQ I was asked to outline why they were successful at winning the Illinois Governor's Pollution Prevention Award at their Schaumberg IL electronics assembly plant. For the record, this award was made public and contains no proprietary information. The full paper was presented at an AIChE symposium on Quality and Pollution Prevention that I chaired in 1994 [2]. A brief overview is presented in Sect. 9.12.

9.12 Pollution Prevention and Waste Minimization: A Quality Challenge at Motorola [8]

Without going into too long a story, I found their method of quality management which revolved around the six-sigma approach treating whatever problem or process they wanted to improve as a defect. For my own interest I calculated six-sigma to be approximately one defect in 3.4 million parts. This is the prime reason their products are essentially defect-free. In this instance, the company chose pollution as a defect and set out to reduce or eliminate solvents from one of their manufacturing processes. According to the award, they successfully modified a particular soldering procedure that had used organic solvents, greatly reducing emissions.

At that time Motorola Safety, Health and Industrial Hygiene group adopted a new corporate wide goal entitled—"Going Beyond Compliance." These solvents were part of the company's air permit and although they were in compliance, they chose to target pollution as a *defect.*

This they did so successfully, that they won the Illinois Governor's P2 award by dropping below the minimum level required for reporting to EPA. In addition, the capital costs for the modifications were recovered within less than 1.5 years, due to savings in solvent costs, operations and maintenance and other elements. Nearly 100 tons/year volatile organic material (VOM emissions) was eliminated through process changes. (And yes, that means 100 tons/year of solvent did not need to be purchased!) Note that this was a permanent change in not only both emission reduction and potential personal exposure but also in nonrenewable resource reduction, as required by SE principles. Needless to say, these modifications were monitored and shown not to reduce the six-sigma production requirements. And of course with such short payback period, the savings accrue right back to the bottom line ... forever. Pollution prevention can pay if done correctly and lead to sustainable operations. Fig. 9.3 depicts the generic quality program model I developed, that I found applied to this specific project. The four corners of hazardous materials management were developed by me and my colleague Richard T. Cartwright, to describe a systematic approach to all manner of Environment, Safety and Occupational Health–based activities. I found it fit well with Motorola's Corporate ESIH Policy "Going Beyond Compliance."

There are three parts to this generic quality model:

1. Central—Quality Management Problem Solving Algorithm—TQM or equivalent.
2. On the Left—Technical System—Specific Body of Knowledge.
3. On the Right—Technical System—Specific Training for all Team Members.

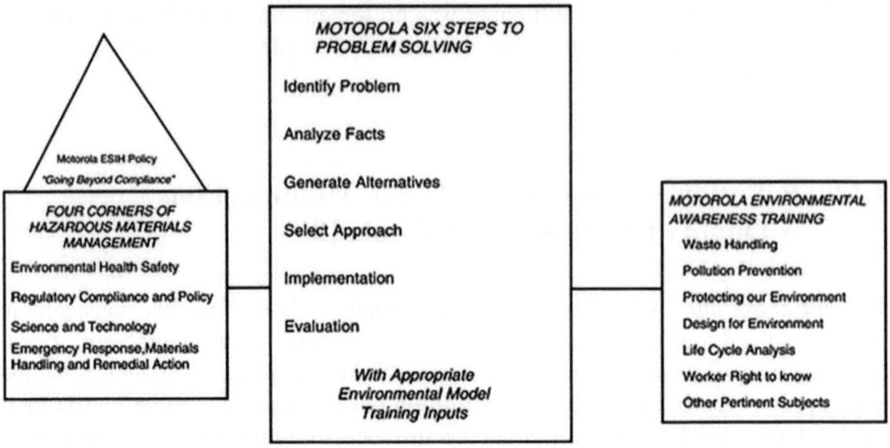

Fig. 9.3 A functional generic quality management system (Perl 1994).

Basically at the core of this approach is your organizations quality management model. Then, plug in your technical model that provides the team with a body of knowledge specific to the problem at hand. Finally, make sure that team members are trained regarding all elements of their assigned tasks.

9.13 Summary

When it gets right down to it, sustainability engineering design is all about optimization and therefore quality management principles must prevail.

References

1. American Society for Quality (ASQ) Total quality management. http://asq.org/learn-about-quality/total-quality-management/overview/tqm-history.html
2. Perl JP (1994) Pollution prevention, quality programs and training. AIChE, Chicago
3. Rumsfeld D (2013) Rumsfeld's rules. Broadside Books, Northampton
4. Grant EL, Leavenworth RS (1980) Statistical quality control, 5th edn. McGraw-Hill, New York
5. OSHA (1992) Process safety management standard. US Department of Labor. osha.gov
6. Aguayo R (1991) Dr. Deming—the man who taught the Japanese about quality. Deming's forward, Simon and Schuster/New York
7. Perl JP (2000) Fixing dysfunctional remediation teams. AIChE spring conference, Atlanta, 2000, and 2006 Orlando
8. Daiya C, Barton P, Perl J (1994) Pollution prevention and waste minimization—a quality challenge at Motorola. AIChE, Chicago

Additional Resources

Curran MA (1994) Life cycle analysis—a new tool for pollution prevention and quality planning. AIChE, Chicago

Perl JP, Cartwright RT (1992) The certified hazardous materials manager review course—the Chicago experience. HazMat World Magazine, July 1992

Perl JP, Cartwright RT (1990) The four corners of hazardous materials management. HazMat World Magazine

Chapter 10
Government Regulatory Development for Sustainability Engineering

In 1970, things in the USA came to a critical mass regarding the need for environmental regulatory intervention. So many major publicized events took place that the past environmental damage caused by unregulated practices of the industry came home to roost and the federal government was forced to act. The ESOH regulations were needed then and now, but now they are institutionalized. I was a party to discussions regarding the question of the early 1990s that asked how industry would respond to the chemical land disposal bans. The answer, the Pollution Prevention Act of 1990, provided a combination of reporting requirements that industry actually met ahead of schedule. While doing so, this activity started to combine with the nascent green movement in consumerism and led industry to see the value in this new proactive environmental compliance approach that became advertisements, not complaints.

10.1 Is Government Interaction Needed for SE?

Spending the last 25 years officing with attorneys exposed me to the "Law of Unintended Consequences" (LoUC). This is a sort of Heisenberg's Uncertainty Principle as applied to social activities, e.g., passing a law to raise interstate highway speed limits from 55 to unlimited, as Montana did in the mid-1990s (actually the law said safe and prudent, which a Montana State Trooper explained to me over coffee was really 85 mph!). The intent was to recognize and make use of the relative emptiness of the highways, but could not anticipate the severity and frequency of additional accidents as the speed approached the limits of *average* human capability. In addition, no effort was made to re-engineer Montana roads for such speeds, e.g., Germany's internationally famous Autobahn, and finally the effect on gas mileage leading to reduced fuel efficiency was certainly never considered, or thought to be an individual's economic decision. Similarly, some say the federally imposed 55 mph speed limits that arose out of the 1973 Middle East oil embargo of the USA

J. P. Perl, *Sustainability Engineering for Enhanced Process Design and Manufacturing Profitability*, https://doi.org/10.1007/978-3-031-52363-2_10

may have led to increases in accidents caused by people speeding and weaving anyway alongside others as well as lost productivity of truckers spending 15+% more time getting from point a to point b (not to mention falling asleep at the wheel).

Our federal, state, and local legislators must take these matters into consideration when creating legislation. This is no less true of designers employing sustainability engineering methodology. So, should SE be mandated? Will the industry apply it alone, or will a combination of carrot and stick be needed? Ultimately the new mpg fuel mandates on small trucks makes it easier for those living in remote areas like Montana, to feel free to drive their 25-mpg pickup that 20 years ago was 10 mpg or so. The cost of imposed efficiency improvement is minimal when spread across millions of sales, and the reduction in resource use is permanent and in keeping with SE design for automobiles. So, what if anything might be required for SE to gain acceptance? Might some federal leveling be required? What form would that take? Should it just be left to individual states? More to the point, what would be the best? Here are some thoughts to consider and build upon.

10.2 Sustainability Engineering: Sanitary Practice Came First

Regarding sanitary engineering practices, the nineteenth century saw great strides in our understanding of the connection between human and animal waste, potable drinking water, and disease. Here, basic science and insurance actuarial investigations led the way to local government identification and action on this important health matter (Ch1. Ref. 7). In my hometown of Chicago, for instance, the first tunnel was dug just after the Civil War, to draw clean drinking water from several miles offshore of Lake Michigan [1]. This engineering feat provided water to the famous Chicago Water Tower and pumping station on North Michigan Avenue. Both structures survived the 1871 fire as they were not made of wood and remind us of how long ago we recognized the need for sanitary fresh water. The pumping station also supplied "force mains" that brought fresh water to the point of use, for drinking, cooking, or fighting fires.

The connection between the typical intestinal, cholera, typhus, and other diseases that manifest themselves when humans live in concentrated close proximity almost immediately caused advances in sanitary practices to proceed rapidly even ahead of regulations (Sect. 1.7). The case, however, for adverse health effects due to industrial pollution would take much longer as lung and other organ disease can take decades to develop.

The Jane Adams Hull House, also in Chicago, helped create and establish the first Occupational Health Study Group. The mad hatters from breathing poisonous mercury vapor during hat band stiffening production, coal miners' lung from mining dust inhalation, then later asbestosis from mining and handling of asbestos, and a host of others occupational diseases began to be recognized that led to significant changes in industrial practices. Now, prior to manufacturing any new chemical, EPA regulations require a PMN as described in Chap. 5, so we know how to do it safely from the start.

✈Society often shoulders huge medical and unemployment compensation costs long after the offending manufacturing process, so this PMN barrier represents a large, positive SE outcome that is often not measured completely upfront.✈

10.3 Government Regulatory History

So, back to 1970, President Nixon, a staunch conservative, signed the EPA and OSHA into law. Though not mentioned at the time, these two acts were major milestones for sustainability in general as they recognized the past and present industrial practices at the time were anything but sustainable. We now can take for granted, or more importantly demand of our government the right to breathe clean air and drink pollutant-free water. We are in far better shape than our ancestors who lived in cities choking from uncontrolled coal and wood burning exhausts, contaminated water, and misunderstood disease transmission vectors.

Enter sustainability as a new consideration and the question is put forth: Are new regulations needed to accommodate our greater understanding of life that has come under the umbrella of sustainability? The answer may be somewhere between yes and no. A very unlikely, positive unintended consequence is the general public's interest in "all things green." Companies are falling over themselves to tout their green-ness and sell into this market.

Perhaps regulations similar to the various product labeling ones aimed at preventing false claims might be useful, but no real methodology for this has arisen. It probably is best to let industry fend for itself regarding energy-efficiency claims. A well-established method of legal remedy exists for adjudicating claims between vendors and buyers of industrial equipment or services. This question is still open for debate.

In addition, as discussed in Chap. 4, the Energy Star labeling program allows consumers to vote with their feet to purchase energy-efficient appliances. We have mandated MPG rating for automobiles and SEER ratings for HVAC equipment that at least assure energy savings on the order of 25–75% for consumers. So, the last great frontier may be direct regulation of industry similar to the ESOH ones in place since 1970. How to do it is the question, by carrot or stick?

10.4 Some Past Successful Mandated Programs

Some past successful mandated programs include:

1. Federal automobile gasoline(MPG) standards, 1974
2. Heating and air conditioning minimum efficiency

3. Federal facility energy use
   (a) Lighting motion detectors, electric utilization, CFL, and fluorescent
   (b) Energy efficiency of all kinds

10.5 "Sustainability Index" → NEW BIG IDEA →

DOE and EPA already publish studies on efficiency, but a "Sustainability Index," e.g., Table 7.1, similar to Energy Star ratings might help SE designers sort through the wheat and chaff of technology selection.

The following are excerpts applicable to Government Regulatory Development; some are private, some are government.

10.6 German Energy Policy and Solar Energy [2]

Here is an interesting example regarding solar energy, a very important sustainable technology written by Tom Friedman in the New York Times in May 2015. Friedman suggests the USA has missed out to some extent, but the point here is that Germany did this through energy policy and incentives rather than regulations. Germany converted 30% of their electric grid to renewable energy through government incentives to create such energy.

I feel the carrot + incentives approach is the way to go here until a clear path, e.g., 78% versus 95% AFUE can be found for SE industrial equipment. In the meantime, however, the DOE, NREL, and EPA are all doing research that feeds into improved energy use, so perhaps a major push on equipment would be helpful. This would also provide another conduit between academia and industry, providing both new developments and more practically trained engineers and technicians.

10.7 "Obama Clean Power Plan": From epa.gov, November 2015 [3]

This executive order was rescinded in 2019 and replaced by the Affordable Clean Energy Rule. From the USEPA Administrator, June 2019:

> *Today, we are delivering on one of President Trump's core priorities: ensuring the American public has access to affordable, reliable energy in a manner that continues our nation's environmental progress,"* ***said EPA Administrator Andrew Wheeler***. *"Unlike the Clean Power Plan, ACE adheres to the Clean Air Act and gives states the regulatory certainty they need to continue to reduce emissions and provide a dependable, diverse supply of electricity that all Americans can afford. When ACE is fully implemented, we expect to see U.S. power sector CO2 emissions fall by as much as 35 percent below 2005 levels."*

The ACE rule establishes emissions guidelines for states to use when developing plans to limit carbon dioxide (CO2) at their coal-fired power plants. Specifically, ACE identifies heat rate improvements as the best system of emission reduction (BSER) for CO2 from coal-fired power plants, and these improvements can be made at individual facilities. States will have 3 years to submit plans, which is in line with other planning timelines under the Clean Air Act.

Also contained within the rule are new implementing regulations for ACE and future existing-source rules under Clean Air Act Section 111(d). These guidelines will inform states as they set unit-specific standards of performance. For example, states can take a particular source's remaining useful life and other factors into account when establishing a standard of performance for that source.

As of this writing, this rule is still in flux and the reader is suggested to follow EPA rule-making. The Clean Air Act, however, is still very much enforced.—JPP

10.8 Energy Conservation in Commercial Buildings [4]

It is not always possible to completely utilize all manufacturing waste heat throughout a plant. Whenever possible, consideration should be given to providing hot water for nonproduction uses, such as sanitation. Space heating as well can be accomplished with low-quality waste process heat from water during heating seasons. All of this is, of course, subject to standard engineering economics review. It would not always be possible to accomplish this type of heat recovery, but remember these represent considerable long-term savings and deserve SE design attention. In some instances, cooling might even be available. Of course, these systems must be isolated, or buffered from the process, a simple task similar to the millions of vented, gas-fired home-heating systems.

Building energy footprints can often be reduced by exchanging fresh and exhaust air. Not only is occupant health improved but much energy can also be saved. Modern CFL or LED lighting should also be employed wherever possible. Replacement of incandescent as well as HID lighting will also reduce heating loads. An SE-integrated power and manufacturing facility should account for all such needs.

Additional EPA Website Excerpt from EPA Sustainable Materials Management follow

10.9 Thinking Beyond Waste: Sustainable Materials Management[5]

From https://www.epa.gov/smm

EPA is helping change the way our society protects the environment and conserves resources for future generations by thinking beyond recycling, composting and disposal. Building on the familiar concept of Reduce, Reuse, Recycle, the Agency is employing a systemic approach that seeks to reduce materials use and associated environmental impacts over their entire life cycle, called sustainable materials management (SMM). This

starts with extraction of natural resources and material processing through product design and manufacturing, then the product use stage, followed by collection/processing and final end of life (disposal). By examining how materials are used throughout their life cycle, an SMM approach seeks to use materials in the most productive way with an emphasis on using fewer materials and products, reducing toxic chemicals and environmental impacts throughout the material's life cycle and assuring we have sufficient resources to meet today's needs and those of the future. *Data on municipal solid waste generation, recycling and disposal is an important starting point for the full SMM approach.*

10.10 Natural Gas Emissions from Fracking

✈The following, from govexec.com, refers to new regulations re: natural gas emissions arising from fracking. You can see from the arguments that an assumption of CH**4**-enhanced global warming is the driver. The industry perspective is that enough is already being done voluntarily. There is money to be made by increasing capture of escaping CH**4**, but only within ROI boundaries. Perhaps the regulations will help create a level playing field so all Oil and Gas (O&G) companies will compete fairly. Let us also not forget that the recovered CH**4** has enormous energy value or, as we have seen, can be converted into chemical feedstock. Here is a recent review of the issue.✈

10.11 Methane Emissions Reduction Program (ca. 2022–2023)

Partial Excerpt from https://www.epa.gov/inflation-reduction-act/methane-emissions-reduction-program

The Inflation Reduction Act provides new authorities under Section 136 of the Clean Air Act to reduce methane emissions from the petroleum and natural gas sector through the creation of the Methane Emissions Reduction Program. The Methane Emissions Reduction Program includes the following components:

Financial and Technical Assistance*—More than $1 billion to reduce methane emissions from the petroleum and natural gas sector by providing financial and technical assistance for preparing and submitting greenhouse gas reports, monitoring methane emissions, and reducing methane and other greenhouse gas emissions from petroleum and natural gas systems, including improving and deploying equipment to reduce emissions, supporting innovation, permanently shutting in and plugging wells, mitigating health effects in low-income and disadvantaged communities, improving climate resiliency, and supporting environmental restoration. The program specifies that at least $700 million must be used for activities at marginal conventional wells.*

JPP Note: This executive order can be found at https://www.epa.gov/inflation-reduction-act/methane-emissions-reduction-program.

10.12 Renewable Fuel Standard Program

Congress created the renewable fuel standard (RFS) program to reduce greenhouse gas emissions and expand the nation's renewable fuels sector while reducing reliance on imported oil. This program was authorized under the Energy Policy Act of 2005 and expanded under the Energy Independence and Security Act of 2007.

JPP Note: check back at https://www.epa.gov/renewable-fuel-standard-program Ref. [7] for most recent updates to this changing standard.

10.13 Final Rule: Management Standards for Hazardous Waste Pharmaceuticals

EPA has proposed to add hazardous pharmaceutical wastes to the federal universal waste program. If finalized, this addition will facilitate the proper management of hazardous pharmaceutical wastes by streamlining the hazardous waste management requirements for the generators of these wastes. The inclusion of hazardous pharmaceuticals may increase demand for management of pharmaceutical wastes, including collection of pharmaceutical wastes via drug take-back programs. More information is at https://www.epa.gov/hwgenerators/management-hazardous-waste-pharmaceuticals

2023 Rule Summary

JPP Note: EPA's Acting Administrator signed the final rule, titled, "Management Standards for Hazardous Waste Pharmaceuticals and Amendment to the P075 Listing for Nicotine" on December 11, 2018 and it was published in the Federal Register (FR) on February 22, 2019.

Final Rule: Management Standards for Hazardous Waste Pharmaceuticals and Amendment to the P075 Listing for Nicotine found at https://www.epa.gov/hwgenerators/final-rule-management-standards-hazardous-waste-pharmaceuticals-and-amendment-p075

10.14 Government Support or Assistance Programs

Some form of carrot encouragement might be useful to farmers to promote integrated waste treatment and electricity generation. This is a huge problem, particularly, in big agriculture, but can also be linked to water recovery for reuse as well.

Consider the use of an electric-utility model of consumer tax breaks for incorporating certain technical approaches, e.g., heat pumps and hybrids even within the industry as appropriate. Here, the utilities, which often operate as a quasi-regulated entity, are highly incented to avoid building new, costly, and difficult-to-permit electric-generation capacity. On the residential side, they have managed this quite well with the government through new energy-efficiency requirements for new HVAC, large home appliances, e.g., refrigerators, freezers, dryers, and energy star in general, quite successfully.

10.15 Municipal Wastewater Treatment Sustainability

An enormous amount of wastewater is processed daily throughout the country. After treatment most of it is simply discharged into rivers, streams, lakes, and even the ocean. This has a large SE footprint and is greatly amenable to improvement. Here are just some SE elements to consider in your designs involving water:

1. Process all the way to drinking water standards, or at least as input to drinking water plants
2. Process to grey water for industrial use, e.g., cooling
3. Explore irrigation use, toilet flushing, and other non-potable water uses
4. Meet lake and river return quality, and release only when points 1 and 2 above cannot be accomplished
5. End combined sewer outputs where feasible to greatly reduce WWTP load and encourage 1 and 2
6. Finance all with municipal bonds—these actually have net positive return on investment!

The US infrastructure is already beyond its end of life. We are now starting to see the once unthinkable, i.e., collapsing bridges. Recently, in Flint, Michigan, lead-polluted water was accidentally piped into drinking lines in order to save money, severely affecting human health.

Some form of funded mandate to address the long-term state of potable water in the USA may be needed. Presently, there appears to be no such planning, with individual states doing as they please. The damming of rivers and pumping dry of deep aquifers, sometimes affecting multiple States, may need federal attention.

10.16 Industrial Permitting Review and Update

As we move into a mode with enhanced recycling, EPA, through existing and future congressional mandates, will need to consider allowing air permit changes when improvement-based emissions either stay the same or improve, provided that Btu footprint is reduced. It should be noted that EPA has authorized many States across the union to act as its permitting enforcement arm and thus will require additional coordination. The States, however, are highly incented to encourage their industries to reduce pollution of all kinds. California improved air quality in the 1980s but forced out the furniture industry to the southeastern part of the USA, with no net national improvement.

10.17 Permit MSW Sites for Resource Recovery Regulatory Update Needed

We may also need to examine permitting MSW sites to operate integrated power and chemical production facilities. Many already collect bio-methane emanating from garbage, but an enormous quantity of metals and other values should be mined and recovered for reuse. Organic material can be mined and gasified into chemical feedstocks and if done with power generation, will result in considerable further savings. These savings should also include returning the landfills to useful property, arable and otherwise, and new regulations may be required here as well. Like water, these projects should be highly amenable to municipal bonds with NPV return on investments.

10.18 Reformatting Hazardous Waste: Recycling Regulatory Update Needed

Hazardous materials are often used during the manufacture of chemical products. Many of these products do not convert 100% of the feedstock into product and generate hazardous waste. In addition, hazardous waste might also be generated during the normal operations even with 100% feedstock conversion with less than 100% product selectivity.

Facilities such as these operate with permits to account for all of the above. Waste material generated is typically shipped offsite for treatment and disposal. But, as discussed previously, virtually all organic material can be gasified into chemicals, e.g., H_2 and CO which in turn as feedstocks themselves, can be processed into any number of useful products. So, the waste from these manufacturing facilities can become gasifier feedstock for use onsite, or offsite and really need not be handled as hazardous waste at all.

Such onsite, or even offsite, reformatting processes will need regulatory identification and reclassification update to encourage this fundamental SE activity. An additional benefit of reformatting includes:

- Reduced HazMat transportation
  - Safety and cost savings
- Reduced HazMat disposal Cost
  - Safety and cost savings
- Ability to convert savings into useful production equipment
  - Economic justification for purchasing new reformatting equipment

10.19 The Need for Government Research and Development Subsidy for Renewables

In the search for renewable resources, the government has played and should continue to play a role.

The EPA, DOE, USDA, and Small Business Administration, for instance, all provide a wealth of related information and data and through small business research and innovation funding (SBIR) grants. Fundamental research is also going on in industry, but our academic institutions are ideally suited to create or find the next renewable. Much is considered basic research, e.g., the search for methods to process algae to oil and make alcohol from waste agricultural products instead of diverting food crops, to name a few. This time-tested method also produces a highly trained, educated, and competent workforce

Disparate laws exist within the EPA that sometimes set recycling against waste disposal and often result in contrary rules that discourage the safe re-purposing, or reformatting, of such former waste products. Of course, this can be addressed by preventing waste creation in the first place. Also, EPA regulations state that a material is not a waste until the generator declares it so. This fact should always guide proper recycling efforts to prevent "sham recycling," or illegal hazardous waste storage beyond 90 days. But still, there may need to be some adjustments to facility operating permits and to ease the process of modifying them in the face of clear process environmental improvements.

Gas and electric utilities are typically quasi-regulated, for example, with rates dictated by local commerce commissions, i.e., not exactly free-market. Be careful when evaluating technical options that have government subsidies. These may disappear. Our highways, rail and air transport and agriculture are all heavily subsidized in one way or another. This is not bad, but must be factored into economic evaluations.

Residential energy-efficiency improvements, e.g., often have beneficial tax incentives to encourage purchasing more efficient options that will be in operation for 20+ years. Perhaps this can serve as a model for industry as well. This also serves to offset the need for costly, polluting new generating plant construction.

R&D dollars spent here will also reconnect academia and industry, a much-needed event!

10.20 International Trade Agreements

At the heart of SE and sustainability in general is the husbandry of energy and material resources. Tearing down entire rainforests may lead to temporary profits in third and developing worlds, but are clearly unsustainable even by the most conservative standards. As noted in the introduction, in order to shield US manufacturing from unfair competition, NAFTA featured a requirement, never really enforced, that pollution not be exported by US firms sending business to the trade partners.

Whether through product labeling or forced requirements, our trade partners must not be allowed to engage in nonsustainable practices on our behalf. Some carrot or stick approach will need to be used to prevent devastation in foreign lands while making cheap goods. Care must be taken to focus only on overall sustainability. Perhaps a tariff on foreign imports produced in a manner inconsistent with US OSHA and EPA regulations, in spirit, not in letter, might provide some protection to domestic manufacturers who do comply.

✈SE Product Labeling—Another Big Idea✈

Supposedly, the new Trans-Pacific Partnership Agreement (TPP) has similar requirements. (JPP Note - TPP was rejected in 2017, but some form of agreement is needed and may yet reappear.) We should consider an SE product label (Chap. 1) that describes the origin of the product, so the developed world consumers can reject nonsustainable products while not harming third-world consumers themselves.

10.21 USEPA Universal Waste Program [8]

The following is an excerpt from USEPA. This program refers to waste generated including pesticides, rechargeable batteries, and mercury-containing items. These materials are all heavily regulated by EPA as they should be, but when these materials are handled by the original generator, e.g., during a manufacturing process, then accommodation may need to be made similar to considering the material being a hazardous feedstock, as many raw materials are, as opposed to a hazardous waste, which places an additional layer of record keeping burden. Again, past poor performance on the part of industry has led to the present regulatory climate, and this is merely brought up as a checklist item for review of potential SE-enhancing legislation.

EPA Excerpt:

1. INTRODUCTION

As mandated by Subtitle C of the Resource Conservation and Recovery Act (RCRA), EPA promulgated regulations governing the safe management of hazardous waste from the point of generation until the point of final deposition. Within these regulations, EPA developed a streamlined management program for certain hazardous wastes, known as universal wastes. The universal waste program provides an alternative set of regulations that reduce the regulatory burden by allowing longer storage of these wastes and reduced recordkeeping. Universal wastes are:

- *generated in a wide variety of settings, not solely industrial*
- *generated by a vast community*
- *present in significant volumes in nonhazardous management systems*

10.21.1 EPA 2.1 Wastes Subject to the Universal Waste Program

(The information in this document is not by any means a complete representation of EPA's regulations or policies, but is an introduction to the topic used for training purposes.)

The federal universal waste regulations include hazardous waste batteries, mercury-containing equipment, pesticides, and lamps. To be covered under the universal waste program, these items must first be identified as hazardous waste. Items that still have product value and that are still being used are not wastes and, therefore, are not subject to RCRA. In addition, wastes excluded from the RCRA definition of solid or hazardous waste are not subject to the universal waste program. Only material identified as a hazardous waste that meets the definition of battery, mercury-containing equipment, pesticide, or lamp in Part 273 can be managed under the universal waste regulations.

Currently, only these four wastes are covered under the federal program. There is, however, a petition process in Part 273, Subpart G, which allows individuals, as well as states, to petition EPA to add other wastestreams to the universal waste program. Universal wastes that are mixed with hazardous wastes are fully regulated as hazardous waste. However, mixtures of universal waste and CESQG waste or household hazardous waste are subject to the universal waste requirements since these hazardous wastes are not subject to full Subtitle C regulation. The first waste stream covered under the universal waste program consists of hazardous waste batteries. The universal waste regulations define a battery as a device consisting of one or more electrically connected electrochemical cells that are designed to receive, store, and deliver electrical energy (§273.6). Also included in this definition are unbroken batteries from which the electrolyte has been removed. Hazardous waste batteries meeting this definition are subject to the universal waste regulations. However, lead-acid batteries being recycled may be managed either as universal waste or by Part 266, Subpart G, which contains special provisions for lead acid batteries. EPA does not have comprehensive data to make interpretations about whether certain battery types are or are not hazardous, so it is the generator's responsibility to determine if that battery is subject to regulation as a hazardous waste.

The second wastestream included in the universal waste program is comprised of certain hazardous waste pesticides. Unused pesticides will often become banned from use, damaged by temperature extremes, or no longer necessary due to crop changes. The universal waste regulations apply to persons managing pesticides that are part of a recall program or unused pesticides that are collected and managed as part of a waste pesticide collection program (§273.3(a)). A recalled pesticide becomes a waste when two conditions occur: the generator agrees to participate in the recall, and the person conducting the recall decides to discard the pesticide or burn it for energy recovery. An unused pesticide becomes a waste on the date the generator decides to discard it (§273.3(c)).Pesticides managed by farmers in accordance with §262.70, which allows farmers to dispose waste pesticides on their own farm in accordance with the disposal instructions on the pesticide label, are not subject to the universal waste regulations (§273.3(b)).

Another wastestream covered by the universal waste program includes hazardous waste lamps. Lamps often exhibit the toxicity characteristic (TC) for mercury or lead, making them a characteristic hazardous waste when discarded. In 1994, EPA proposed two options for lamp management (59 FR 38288; July 27, 1994), and in 1997, a study was conducted to evaluate mercury emissions from discarded lamps. After responding to comments received on both documents, EPA added hazardous waste lamps to the universal waste regulations. A universal waste lamp is defined as the bulb or tube portion of an electric lighting device (§273.9). Examples of common universal waste electric lamps include, but are not limited to, fluorescent, high intensity discharge, neon, mercury vapor, high pressure sodium, and metal halide lamps. Several states added mercury lamps to their universal waste programs prior to the July 1999 final rule. Therefore, this federal addition made the management requirements consistent with many state programs. The approach also encourages the manufacture of low-mercury lamps in industry.

The most recent wastestream covered by the universal waste program is mercury-containing equipment. Elemental mercury is contained in many types of instruments and devices that are commonly used by electric utilities, municipalities, hospitals, and other industries, as well as households. Such devices may include thermometers, manometers, barometers, relay switches, mercury regulators, meters, pressure gauges, and sprinkler system contacts. EPA had considered including this category in the 1995universal waste rule, but had not done so due to a lack of information about the wastestream. A 1996 petition to add the category to universal waste provided such information and, therefore, EPA decided to include mercury-containing equipment into the federal program. Because mercury thermostats meet the broader definition of mercury-containing equipment, EPA replaced the previously existing sections regulating mercury thermostats with the new requirements for mercury-containing equipment (§§273.13(c) and 273.33(c)). The management requirements in the new section are almost identical, but add provisions for managing mercury that is not in ampules.

✈This really does not cover the industry need for greater latitude in processing hazardous waste generated on site, e.g., amenable to gasification or detoxification reactions.✈

10.22 Existing Non-ESOH Government Regulations Impacting Design

NEW Manufacturing Considerations—Conflict Materials Regulations (Excerpt) [9]:

Companies registered with the US Securities and Exchange Commission (SEC) will have to disclose the use of conflict minerals sourced from the DRC or its neighboring countries in their products under section 1502 of the U.S. Dodd-Frank Wall Street Reform and Consumer Protection Act (PDF). Companies must file their first reports with the SEC by the end of May 2014.

10.23 Energy Policy Act of 2005 [10]

This is a massive Government energy conservation program, 551-page document, located on DOE website as pdf. See also Chap. 4, Sect. 4.1.2.5.

10.24 Incentives to Power Industry Gasification Projects

Using combined (integrated) power and chemical manufacturing, this can provide additional economic incentive, particularly where CO_2 capture is involved.

From EPA website epa.gov (links removed from below):

EPA Energy and Environment Guide to Action
Policy Considerations for Combined Heat and Power
Policy Description and Objective
EPA Summary

Several policy opportunities exist for states to support greater use of combined heat and power (CHP). CHP, also known as cogeneration, is the simultaneous production of electricity and heat from a single fuel source (EPA 2007, 2014b). CHP provides an alternative to purchasing electricity from the local utility and burning fuel in an onsite furnace or boiler to produce thermal energy (such as steam). An industrial, institutional, or commercial facility can instead use CHP to provide both electric and thermal energy services in one energy-efficient step by capturing and using surplus heat that would otherwise be wasted when generating electricity. Due to the increased system efficiency, the CHP system produces the same amount of energy while requiring less fuel; it also produces lower emissions overall than equivalent, separate heat and power systems.

Optimally designed CHP offers environmental and climate change benefits to states, communities, businesses, and institutions through increased energy efficiency and reduced fuel consumption. Other benefits include improved fuel efficiency, enhanced resiliency, and more reliable power and thermal energy supplies. These reliability and resiliency benefits bolster business competitiveness, the energy infrastructure, and energy security (EPA 2013b). These benefits are enhanced when a CHP system is used in a district energy system or a microgrid.

Recognizing CHP market growth benefits and barriers, President Obama issued Executive Order 13624, "Accelerating Investment in Industrial Energy Efficiency," in August 2012 (White House 2012). The executive order called for the installation of 40 gigawatts (GW) of new, cost-effective industrial CHP nationwide by the end of 2020. It specifically recognizes the lack of a single solution to addressing market barriers and looks for support through a variety of approaches. These include encouraging private sector investment by setting goals and highlighting investment benefits, improving coordination at the federal level, encouraging federal agencies to partner with and support states, and identifying investment models beneficial to the multiple stakeholders involved.

JPP Note —Here is a current link to the USEPA CHP Guidance: *https://www.epa.gov/sites/default/files/2017-06/documents/guide_action_chapter6.pdf*

This is quite important to SE design as CHP combined with chemical manufacturing of disparate products can lead to enhanced economic advantages. Also some regulatory means to process hazardous waste onsite to produce energy or other chemicals is badly needed.

10.25 Summary

Federal regulations, e.g., USEPA, USDA, FDA, OSHA, are the only way of bringing issues of long-term sustainability to the table in a fair and equitable manner. Without regulations, the temptation to slip back to short-term financial gains, indeed, windfalls is far too great. The weather cycles involving real-world sustainability may take too long to be self-limiting. As we learn more over time, the regulations can be modified up or down to fit the facts, rather than emotions. This has happened with lead and mercury, both which have been found to be even more dangerous than previously thought and PCB and dioxins, which seem now to be not quite as great a threat as once thought—all science based.

Drilling for groundwater in California and Florida has caused severe subsidence, despite being in plain sight of two oceans. Desalination through SE methodologies may prove a solution. Regulations may be needed to husband this vanishing resource.

This was indeed a cursory look at this topic. Try to avoid politics here at all costs. Fact-based decision-making should guide our legislators to choose a carrot or stick approach. Public safety and health must guide. The health of US industry need not suffer if new regulations are needed, and may need protection if other countries do not participate. The ESOH regulations of 1970, imposed on an industry incapable of self-regulation at the time, led to an enormously successful increase in quality of life as well as US industrial profitability as it continues to thrive in an "everything green" market-driven economy.

References

1. Chicago drinking water tunnels. CityofChicago.org
2. Friedman TL (2015) German energy incentives. http://nyti.ms/1bxcbxT, OP-ED May 6, 2015
3. Clean Power. http://www.epa.gov/cleanpowerplan/clean-power-plan-existing-power-plants
4. ANSI/ASHRAE/IESNA Standard 90.1 (2013) Energy standards for buildings, except low-rise residential buildings
5. Thinking beyond waste - Sustainable Materials Management From https://www.epa.gov/smm https://www.epa.gov/smm USEPA
6. USEPA From - https://www.epa.gov/inflation-reduction-act/methane-emissions-reduction-program
7. EPA Fuels Standard. http://www.epa.gov/renewable-fuel-standard-program
8. EPA. Universal waste program—various web links. www.epa.gov
9. SEC Conflict Minerals. https://www.sec.gov/News/Article/Detail/Article/1365171562058
10. Energy Policy Act of 2005, US Dept of Energy., energy.gov/sites/prod/files/2013/10/F3/EPACT 2005–2005.pdf

Additional Resources

EPA (2015) Energy and environment guide to action: chapter 6. epa.gov
Occupational Health at Jane Adams. www.hullhousemuseum.org
Doctor John Snow blames water pollution for cholera epidemic. http://www.ph.ucla.edu/epi/snow/fatherofepidemiology.html

Chapter 11
Sustainability Engineering in Various Engineering Disciplines and Industry Segments: Challenges and Opportunities

11.1 Engineering Disciplines

Most of the examples in this book have been drawn from the chemical process industry (CPI). The importance of activities in the other disciplines, however, is equally important. I have highlighted examples where great improvements in process efficiency and improvement have been made and are needed. Clearly, all the activities listed below will need to fall under strict SE criteria as well. At least ranking them would be a good start as unacceptable, good, better, or best.

Creation of a modern CPI plant or refinery has at its core ChE process and product design. The finished product, however, is in no small way, a result of all engineering discipline efforts. Once designed, the plant has to be constructed by CE and SE, equipment built and supplied by ME, laid out by IE, controlled by EE, and operated by ChE and EnvE. And let us not forget AgE designers in big agriculture. Of course, there are overlaps, but you get the picture.

Roughly speaking, in the CPI for large-scale projects, architects and engineers (A&E) are involved up to the detailed design point. The A&E functionally passes off to an engineering, procurement, and construction (EPC) firm capable of taking those process designs and converting into issue for construction (IFC) packages and in some instances, delivering a completed commissioned and started project. With so many engineering disciplines, what follows is necessarily of a cursory nature, but this will jump-start the conversation and team building as well.

J. P. Perl, *Sustainability Engineering for Enhanced Process Design and Manufacturing Profitability*, https://doi.org/10.1007/978-3-031-52363-2_11

11.2 Agricultural Engineering (AgE)

The USA is a very, very large producer of food for our own as well as external consumption. This discipline may well be representative of the greatest SE example. We do not have any land to waste, literally and figuratively! Preservation of topsoil is a giant AgE SE outreach. Also, the waste involved in harvesting, shucking, etc. can be converted biologically into fuels or gasified as well. There is much work going on here and lots more to do. Fertilizer improvements of the type I mentioned previously e.g., Chap. 5, Sect. 12, continue to make crop efficiency improvements for example.

But the enormous amount of animal waste produced on large meat, pork, and poultry farms needs to be addressed with an appropriate engineering effort. There is quite a bit of energy and nutrients that should be reincorporated into the soil system from waste recovery, and certainly not contaminating streams. The whole concept of large-scale farming is amenable to engineering improvements as well. So, perhaps topics such as crop rotation and pesticide use will be added to their tasks. This discipline is ripe for positive SE contributions in all these areas.

11.2.1 SE Applied to Agriculture

Here, the farmer tills, plants, tends, and harvests, then repeats annually. Sustainability, however, would draw a wider circle to examine the long-term effect of fertilizers and pesticides as well as GMO, water use, soil erosion, etc., and may require additional regulations not to simply tax, but to create a level playing field among farmers, which are increasingly large conglomerates, including plant and animal production. Other AG sustainability issues include overuse of antibiotics, water pollution, and spread of infectious diseases associated with modern animal farming. Antibiotics and vaccines are a science-based approach. The need for these, however, can be engineered out through sound SE principles, increasing efficiency while saving money as well as earning new customers willing to pay a premium for such agricultural products. Consumer interest in natural AG products of all kinds continues to translate into premium pricing and therefore greater profits for small and large agri-businesses.

Again, as in the case of global climate change, the causes, often subject to emotion and politics, are far less important than simply engineering systems to accommodate the factual climate change effect on modern processes. Sea level rise from polar ice cap melting must be incorporated into coastal designs of community and industry, regardless of the cause. You cannot grow pigs that have become antibiotic-resistant. We also do not want to introduce our own form of disease in humans, e.g., avian- and swine-based caused by poor, outdated, and uninformed agricultural practices. These are practical concerns requiring overarching supervision connecting cause with effect. So, SE will need to encompass all the above concerns in farming. We remember that past cultures created deserts through poor agricultural practices.

Agriculture, particularly small farmers, may need government tax or large pooled insurance-type incentives to work toward SE practices. Loss of topsoil remains an issue needing continuous improvement.

A good example of the slippery slope of unmanaged, unscientific agricultural practices of the early twentieth century is the Dust Bowl of the 1930s that our parents and grandparents can remember, and the positive Department of Agriculture response (Thanks Ken Burns for reminding us all of this!) Good, science-based SE design is all it takes.

11.2.2 Compost Type Examples: Nothing Is a Waste Until You Say It Is!

Excess animal farm waste also has energy values that if harvested could solve the bio-contamination issues, use as evaporative heat source to purify wastewater and make electricity. Remember, it is not a waste until you say it is as long as it is handled properly, and this type of processing also leads to reduced waste disposal costs, hence larger profits on top of reduced pollution from uncontrolled runoff.

> Excerpt from dairynet.com [1]:
>
> *How does animal waste-to-energy work?*
>
> *Dairy cow manure is the resource to generate renewable energy. The manure is collected and heated, creating the natural byproduct of methane gas. That biogas is the fuel used to power the generators and create electricity.*
>
> *Renewable energy & environmental stewardship*
>
> *The manure digester process has additional environmental benefits, reducing animal waste problems associated with manure disposal on farms. The odor is greatly reduced, and weed seeds and pathogens are killed during the digestion process, thus lessening the need for herbicides and pesticides on the farm. Another useful byproduct is bedding that can be used in the dairy, or sold by the farmer.*
>
> *Excerpt From: AustinGroundtoGround.org* [2]
>
> *Coffee grounds and other restaurant waste recycling examples—from Lindsay Razzaz, Austin TX Austin*
>
> *Used coffee grounds are an immediate and slow release nitrogen source that plants love! They also contain phosphorus, potassium, magnesium, calcium & other trace minerals not found in most commercial fertilizers. Using grounds to fertilize reduces the chance of "burning" plants, polluting groundwater, or killing soil organisms that are part of a healthy ecosystem. Spent coffee grounds are slightly acidic (pH 6.2-6.8)—perfect for Austin's alkaline soils! Plus, the organic matter helps our soils to retain moisture.*
>
> - *Add them to your compost or vermiculture bin.*
> - *Add grounds directly to the soil. Dig into the top couple inches or just sprinkle on top.*
> - *Incorporate up to 25 % the volume of your soil to improve soil structure in the short and long term!*

✈Sure, this would not get much chemical industry use, but AG is a huge business itself. But it is good to know, right up there with recycling deep restaurant frying fats and oils, which is a huge industrial process, making biodiesel, etc.✈

See also Chap. 8, Sect. 8.2.16 "Turning Farm Manure into Renewable Natural Gas," for an industrial-scale AG example.

11.2.3 Agriculture and Education in the USA

The USA came into being as a largely agrarian society on the cusp of a nascent Industrial Revolution. Congress realized the importance of enhancing our understanding of agriculture through education of the farm community and passed this significant law during the height of the US Civil War. To this day, the country benefits from the R&D coming forth from these numerous State institutions.

Excerpt from the US National Archives:

I The Morrill Act of 1862

Passed on July 2, 1862, this act made it possible for states to establish public colleges funded by the development or sale of associated federal land grants. Over 10 million acres provided by these grants were expropriated from tribal lands of Native communities. The new land-grant institutions, which emphasized agriculture and mechanic arts, opened opportunities to thousands of farmers and working people previously excluded from higher education.

Sponsored by Senator Justin Morrill of Vermont, the "Act Donating public lands to the several States and [Territories] which may provide colleges for the benefit of agriculture and the Mechanic arts" marked the first federal aid to higher education. But the government's recognition of its obligation to provide schools for its future citizens dates from the beginning of the republic. In the Northwest Ordinance of 1787, the Continental Congress wrote, "Knowledge, being necessary to good government and the happiness of mankind, schools and the means of education shall forever be encouraged." With this ordinance, Congress established a precedent for the support of public education that would grow to substantial commitments in later years.

11.3 Biomedical Engineering (BiomedE)

I have always felt this was a logical offshoot of ChE. The organs of the human body, lungs, skin, stomach, intestines, etc., can all be mathematically modeled by ChE process. The stomach is a bioreactor converting food into energy; the lungs, kidneys, and liver are mass transfer devices. The heart is nature's perfect fluid pump. Kidney dialysis, newer and more efficient artificial hearts, along with the required mechanical engineering, are a few examples of improvements coming from this discipline (with a big tip of the hat to electrical and computer engineers who keep it all connected!).

11.4 Chemical and Biological Engineering (ChE and BioE)

Turning large quantities of raw materials, safely and economically, into finished products is the principal activity here. Any conversion of chemicals on a large scale into finished products is really the purview of ChE. Crossovers with biological, environmental, and agricultural, themselves logical subsets of chemical engineering, are bound to happen. More recently, Sustainability has become a buzzword outreach for ChE, and the bulk of this book addresses the CPI and its unique ability to enhance power conversion efficiency and efficacy through capture and reuse of both high- and low-quality waste heat from power generation and exothermic chemical processes. The power portion is obviously mechanical and EE and remember, CE and SE will build the whole thing! In some schools, ChE is stand-alone; in others, biological engineering was added, as so many advances are now based more on biological, living processes in addition to chemical ones, but the concepts of mass production are largely the same for each making a natural coalescence of the two. ChE and BioE will make great inroads toward SE as they focus on renewables.

11.5 Civil and Structural Engineering (CE, SE)

One of my buddies once told me civil engineers are not happy unless they are "Moving dirt, pouring concrete and hanging steel!" The largest branch of engineering, we see CE examples every day. Our buildings, roads, highways, airports, shipping ports, railroads, and bridges to name a few are all civil engineering projects that come to life. Structural engineering (StE) is a sort of stand-alone branch of CE to provide the necessary safety for all CE projects, particularly since most are built to last a lifetime and need to support design loads, or else! And lasting a lifetime is definitely SE.

11.5.1 Wastewater

Wastewater generation and treatment is enormous in the USA, and civil and environmental engineers both work on these very large-scale projects. We are finally getting around to sterilizing this stuff, but as the Los Angeles Mayor recently lamented, "…why are we not reusing this for agriculture or perhaps even drinking water purposes?" We also at one time had lower purity "city water" (non-potable) that was used for process cooling and other non-sanitary purposes, and perhaps the CE and EnvE might bring this back or just treat wastewater to drinking water standards for purest level of SE. This might require regulatory updates as discussed in Chap. 10.

11.5.2 Road Building: The End of Asphalt?

It may not be obvious, but the asphalt used to cover our highways was really once an unwanted waste by-product of refining of crude/lube oil. If a way is ever found to completely replace petroleum-based fuels in the future with electricity, hydrogen, biofuels, natural gas, or similar transportation fuels, then some suitable substance will need to replace asphalt, or it will need to be artificially manufactured just like any other chemical product. This, in turn, would need to be factored into any switch of transportation fuels. Alternatively, roads could be made out of uncovered concrete and perhaps a way might be found to improve crack resistance, extend highway life. The point is the SE designer will need to evaluate a broader range of cause-and-effect relationships related to their designs. These new engineering challenges will lead to better and less expensive ways of doing things, but not without R&D to back them up. Perfectly suited concepts to introduce to academia for research-based improvement.

11.5.3 Concrete Is Ubiquitous: Recycling It Is Not Obvious Here—Economics Drives Decisions

Excerpt taken from wbcsdcement.org [3]:

> *Recovering concrete has two main advantages: (1) it reduces the use of virgin aggregate and its associated environmental costs of exploitation and transportation; and (2) it reduces unnecessary landfill of valuable materials that can be recovered and redeployed. Recovering concrete, however, has no appreciable impact on reducing greenhouse gas emissions. In the product life cycle of concrete, the main source of carbon emissions is the cement production process (cement is added to aggregates to make concrete). The cement content in concrete cannot be viably separated and reused or recycled into new cement, and thus carbon reductions cannot be achieved by recycling concrete.*

Reference wbcsdcement.org and ashgrove.com Cement Company.

11.6 Computer Science and Engineering

Advances in chip design and power footprint are continuous and have much room for improvement. Millions of machines are on or in sleep mode constantly. Server farms must be air-conditioned, which should motivate the development of new computational materials. In the meantime, some of these buildings at least capture the heat within the HVAC system itself, a perfect application for a heat pump.

11.6.1 Process Instrumentation and Control

No process can run without instrumentation and control. More emphasis will be needed on practice-based application alongside the mostly theory-based approach. Some departments have dropped this from the curriculum; so, ChE process design education will need to pick this up. Integration with offsite power as well as design modeling programs will also inform instrument product development. The offsite power integration challenge will require sophisticated, controlled interconnection between plant and grid. See also Sect. 8.5.

11.7 Electrical and Electronic Engineering (EE)

Well, now that you think about it, EE is almost as ubiquitous as CE. Computers and all electronics consume small amounts of electricity, but when multiplied by 100s of millions turn out to be a large number. As discussed in efficiency of all things Chap. 4, the conversion of electricity to useful work by electronics is quite low. As computer chip component spacing gets smaller, newer heat dissipation methods will be needed as well as new methods for creating transistors that are more efficient in the first place. What to do with this waste heat? Chillers are required for server farms or even offices with large numbers of machines. Electrical engineers along with mechanical engineers generate electricity on a massive scale and will have much to contribute to SE in future integrated power and chemical production facilities. Power electronics for LEDs, utility interconnection issues for solar-, wind-, and facility-generated power for export, e.g., power factor, surge, timing, control, and other elements are the domain of electrical engineering.

11.7.1 Incandescent Lighting: The End of Waste Heat?

Thomas Edison's trial and error invention of the filament-based electric light bulb is the stuff of engineering legend. Edison was well aware of the waste heat (Chap. 4).

The Edison bulb, of course, is largely wasteful of electricity as less than 3% is converted into light. But that early and still ubiquitous illumination method served as a critical bridge to the future. The next breakthrough in lighting was the fluorescent bulb ca 1920s. This was followed by halogen lighting ca 1959/60 and the light-emitting diode (LED).

Excerpt from DOE.gov:

They (LED) are also the most efficient lights on the market. Also called ***luminous efficacy****, a light bulb's efficiency is a measure of emitted light (lumens) divided by power it draws (watts). A light bulb that is 100 percent efficient at converting energy into light would have an efficacy of 683 Lm/W. To put this in context, a 60- to 100-watt incandescent bulb has an* *efficacy of 15 Lm/W, an equivalent CFL has an efficacy of 73 Lm/W, and contemporary LED-based replacement bulbs on the market range from 70-120 Lm/W with an average efficacy of 85 Lm/W.*

From DOE, we see that the LED bulb is 85/15 or nearly six times more efficient in converting electricity into light than incandescent ones, but still only converts 12.4% electricity into light. Where does the lost energy go? The LED itself is relatively cool so look to improvements in associated power electronics to eventually solve that dilemma or examine the electromagnetic spectrum output to learn if we just cannot "see the light," e.g., UV and IR from the LED. New government regulations are in the works requiring standby power efficiency improvements, e.g., microwave oven control and display electronics.

Of note: The LoUC—the CFL or LED cannot be used with conventional dimmers. These can cause either bulb or dimmer damage. They also exhibit a reduced electrical power factor (PF) which will increase effective loading on power plants. At present, this represents lost revenue for utilities but look to smart meters to perhaps recover this. I only bring this pedestrian example up here in this CPI-based book as it provides a prime example of the need for more thorough technical review of options that SE can help bring to the fore. And do not throw away your old incandescents just yet, they can be used in cold climates as a heat source during the winter, thereby extending the LED or CFL life and often better replicating sunlight during that often dreary season. Dermatologists also use them in light therapy.

11.7.2 Photovoltaics

Nature through photosynthesis reminds us of the immense supply of free energy hitting the earth's surface daily. Advances are being made constantly at increasing efficiency and efficacy of converting this energy into electricity for direct or stored usage.

Lighting will go from Incandescent→Fluorescent→LED in the future. Improved power electronics will be needed to support advanced lighting application reliability and efficiency. The one exception of course occurs where color balancing is required.

On average, even after passing through hundreds of kilometers of air on a clear day, solar radiation reaches earth with more than enough energy in a single square meter to illuminate five 60-W lightbulbs, or 3000 kW-h/month, if all the sunlight could be captured and converted to electricity [4]. This will greatly impact residential but not as much industrial. However, this can greatly reduce existing load, leaving plenty of electric generation capacity for industry, and this has been a very successful task taken up by government and utilities to both reduce energy footprint and avoid building expensive new generation capacity. This goes for residential efficiency improvement of all kinds. According to eia.gov about 3.4% of electric power is coming from photovoltaics, with tremendous room for growth. There are ripe opportunities for urban community organizers to capitalize on this!

11.8 Environmental Engineering (EnvE)

I am looking to these good people to foster greater conversion of MSW not just to energy, but to resource recovery. Around 28% of MSW is organic material with heating values of ca 8000-10,000 btu/hr affording tremendous opportunity for energy and chemical reformatting. Many plants utilizing solvents operate thermal oxidizers to reduce emissions, and energy here is often wasted. As noted elsewhere, this is a great US business opportunity (The two Rs in RCRA are Resource Recovery.) Also, while you are at it, take a swing at algae to oil before the chemicals beat you to it! Also, too much recycling of electronics is going offshore with some being improperly disposed of. We may need regulations to prohibit that with the positive effect being to create more jobs here while polluting less elsewhere. The fundamentals of chemical engineering apply to this discipline regarding materials, energy, and reactions. Some 40–50 years ago, environmental was on the national forefront and so split off on its own. This was probably a good idea considering little interest in such matters within the ChE community at that time. Today, it is common to find environmental within civil engineering departments.

11.8.1 Construction Waste

As we move toward a fully sustainable society, construction waste will need to be incorporated into the mix as feedstocks. As I noted earlier, in CE, asphalt, concrete, wood, plastic, metal conduit, and plumbing and HVAC piping should never enter the landfill. Large municipalities can use this as a revenue stream. The waste management entities can easily set up to capture these materials and parse them and resell, or even become wholly captive industries similar to paper. Essentially nothing should go permanently into landfills. Landfills should serve only as temporary storage or as chemical reactors for gasifying bio-waste into methane, hydrogen, and other materials for either onsite fuel or integrated chemical production.

I was on my morning walk today and saw a home in my neighborhood getting a new roof. A large crew was busily stripping the old shingles off the roof and tossing them into a 20-yard roll-off box, undoubtedly heading to an MSW landfill! Sure, it is not typically hazardous, but what a waste of money and resources. The many thousands of roofers doing this nationally should be paid for this material, or at least it should be hauled off for free by companies smart enough to develop recycling/repurposing technology.

Here are a few recyclers I found on the web [5]. Claims include using the shingles as admix to pavement asphalt, but from my experience, this needs to be done carefully and selectively due to strict DOT standards and specifications for roadbed.

1. http://asphaltshinglegrinding.com
2. http://www.asphaltroofing.org/asphalt-shingle-recycling-faqs

The second website is an organizational one with numerous references.

Please make sure your next plant turnaround or rebuild project does not ignore asphalt, concrete, steel, and other construction debris recycling options. You would be surprised how often this gets ignored. And yes, this saves money over landfill costs, and best of all is handled by others taking workload off your staff. But I promise it would not get done unless it is a mandated checklist item.

11.8.2 ESOH Plant Operations

Often the domain of environmental engineering, this offers a key entryway into SE design. Remember the quality-team approach wins the day so EnvE must help ChE stick to the SE objective in plant design and operations. There is plenty of opportunity for improvement here in working toward SE.

11.8.3 SE Design Goal of Zero MSW: Conversion of Garbage to Useful Energy and Materials

→A KEY SE COMPONENT!→

It has long been known that municipal solid was (MSW) contains high Btu content suitable for fuel, but the presence of toxic metals, e.g., lead and mercury require special air and water pollution treatment. Gasification of MSW into syngas, i.e., CO and H_2, can be done under less harsh conditions and actually allow the recovery of metal values as well as glass for recycle and reuse. Fully understanding this can be incorporated into integrated power and chemical production plants as well as city and regional planning efforts. Gasification itself is a very mature technology that can easily convert hydrocarbons of all types into H_2 and CO. The H_2–CO ratio can be varied and combined with other atoms to make, e.g., NH_3 and then fertilizers and fuels. This has been discussed in some detail starting with Chap. 3. Before the advent of natural gas, town gas plants manufactured medium Btu pyrolysis gas, a modified form of gasification that created tremendous quantities of hazardous waste. This waste was commonly buried onsite in surrounding soils that we are still cleaning up and, which even today could be used as chemical feedstocks if regulations were adjusted.

11.8.3.1 Basic Information About Landfill Gas: USEPA

Methane Emissions from Landfills Municipal solid waste (MSW) landfills are the third-largest source of human-related methane emissions in the United States, accounting for approximately ***14.3 percent of these emissions in 2021****.* *(Aug 3, 2023)*

(Excerpt from https://www.epa.gov/lmop/basic-information-about-landfill-gas)

Many landfill operators already collect off-gas. It can be dewatered and used in gas engines for power generation. It is often flared, but can also be upgraded and used for pipeline heating and cooking gas. The best option would be to avoid MSW completely as described in earlier sections.

11.9 Industrial Engineering (IE)

For some reason, just as I was entering graduate school in the late 1970s, this discipline started to disappear from campuses. But this is the discipline that knows how to organize people and manufacturing on a large scale without which there is no way to optimize manufacturing of any kind! Plant and production layout by IE can lead to reduced pumping power as well as plant footprint and attendant energy use. How the plant floor is organized and how raw materials and product moves and is handled improves manpower and energy efficiency.

IE is slowly returning and must be reborn if we are to staunch the loss in domestic manufacturing, and optimize what we still have. IE may be the original quality management implementer. As our plants become more automated, IE will lead the way in keeping and bringing back a good deal of our domestic manufacturing capacity.

Do it right up front or lose operating and maintenance (O&M) $ in the long run!

11.10 Mechanical, Materials, and Aerospace Engineering (MMAE)

MMAE also contains automotive engineering. The conversion of energy into power in automobiles is not very high. The advent of the modern diesel locomotive, in the 1930s, created essentially the first hybrid electric and combustion motive force. "Electromotive" propulsion made rail a very inexpensive mode of shipping as well as transportation. The advent of the modern jet engine and its power analog the turbine generator have also both led to remarkable increases in energy conversion efficiency. Over the last 10 years or so, the hybrid and full rechargeable automobiles have appeared on the market. My hybrid 2005 Prius, for example, still gets nearly 47 mpg year-round, nearly double my last car and my 2019 is up to 50+ mpg. Pure electrics can get in excess of 100 miles per gallon – electric equivalent (MPGe), an EPA-/industry-calculated electric equivalent to gasoline. For the record, fully loaded jet aircraft attain relatively high seat-miles/gallon of fuel. MMAE also designs much of the components that comprise modern HVAC systems with ASHRAE, a largely MMAE/EE hybrid. HVAC SEER ratings continue to climb and selection of these up front also assure long term savings. Building ventilation as well.

11.10.1 Central District Heating: The Return to What Used to Work [6]

In 1882, when Thomas Edison built the world's first commercial power plant in Manhattan, he sold its waste steam (low-quality turbine extraction steam) exiting the generators to heat nearby buildings. Methods of using this same approach in integrated chemical and other manufacturing production is the point of this book.

For further details, refer to the weblink, districtenergy.org/blog/2014/02/25/waste-heat-is-free-energy-so-why-arent-we-using-it/.

Edison's first major power station was a combined heat and power unit but did not use heat for chemical production, only for neighboring, "district" heating.

Early Steam Adopter—P.T. Barnum used steam heating in his NYC Museum in 1864. For safety purposes, he put the boiler in the basement and kept open flames away from his popular and expensive exhibition space. (JPP—Unfortunately, he was unsuccessful in preventing fires started by his next door neighbor!.)

11.10.2 Geothermal Energy: New SE Adjunct [4]

To access this, the ME needs to work with geotechnical types to bore into the earth and insert their heat transfer equipment. More work will need to be done on creation of very long life subsurface equipment. Residential heat pump systems are replacing HVAC for new homes as well as replacement and retrofits. Large-scale industrial application is possible, but requires careful subsurface design to match above ground process loads to subterranean heat transfer rates and capacities. Huge efficiency improvements over classic HVAC are available here. Industrial heat pumps work on nearly the exact same principle.

> Excerpt from the National Academies of Science and Engineering re: geothermal applications:
>
> *Finally, the heat of Earth's molten interior, itself largely the result of the nuclear decay of radioactive elements, provides geothermal energy. At present, it is chiefly used in only a few places, such as California and Iceland, where proximity to high temperature geothermal fields makes it practical.*

11.11 Metallurgical and Materials Engineering (MetE)

The reason our structures and machines all stay together, do not rust, and our buildings and bridges do not collapse is due to the continuous advances of this discipline. When any designer picks a metal for use in an application, they need to know not just the physical properties, but how long it will retain them. Of course, this applies to all engineering materials as well as metals. To me it gets down to the functionality, reliability, and of course cost. But will these metals be available forever? What about recycling and refining? How about recovering small amounts of gold plating from huge amounts of discarded printed circuit boards? New battery metals provide additional challenges here as well. MetE has some very interesting SE challenge to chew on!

11.12 Nuclear Engineering

The only other original source of energy on earth's surface is found in more local nuclear reactions, where atoms of radioactive elements such as uranium split apart into smaller atoms and liberate energy in the process. Harnessed as heat, the released energy boils water, producing steam that turns turbines, thereby being converted to mechanical energy that generates electricity. Nuclear energy currently provides 18% of total electricity generation in the USA [4]. Great possibilities still exist for fusion energy reactors, the energy of the sun, if containment issues can be solved.

11.13 Industry Segments: Where the Disciplines Are—Challenges and Opportunities

This will not be a complete listing of all potential industry, but here goes. Please send those cards, letters, and emails to update the next edition. In a consumer-oriented society, sustainability can be enhanced by recycling and where possible, repurposing. There is already a very large amount of recycling going on here but far from the nearly 100% it could be. Keep in mind also that many industries do a poor job of recycling MSW and metals, etc. True, they do not represent a large percentage of material flow at any given facility, but everything matters in the greater scheme of things, so please make sure MSW recycling is a key element at your facility.

Many facilities I have visited have operated for years under permits allowing waste disposal. Be they air or water releases, they should be reviewed periodically for process improvement. Particularly where SE is involved we must remember that most everything is reusable somewhere, often as a feedstock, and an ROI of 2–3 years is not unheard of, but as usual will take review and engineered solutions.

11.13.1 Agriculture

I am not sure why folks do not always think about agriculture as an industry, but it sure is one of our most important ones. Whether small farms or large agri-business, it keeps America fed along with large parts of the world through exporting. Some of the more interesting sustainability issues are as follows. (See also Sect. 11.2.1)

11.13.1.1 Crop Production: Pesticide Use, Soil Erosion, and Bee Loss

Adverse effects are important to minimize as they relate to the interconnectedness of life on earth as well as the ability to sustain crop production to feed an enormous and growing global population. Modern AG methods have allowed this overextension of population support and the loss of any of the key naturally occurring components could resonate with severe consequences within a short period of time. Unlike oil, we have no viable backup for our food production system.

11.13.1.2 Animal Farming: Antibiotics, Animal Waste, Meat/ Vegetable Balance

Dairy, poultry, beef, pork, all produce vast quantities of organic waste that are perfectly applicable to the same type of SE processing used for MSW. There is no reason any of these wastes need ever migrate into rivers, streams, or cause infections among the herd. Antibiotic overuse is another sustainability issue as it could lead to disease-resistant viruses and bacteria as happened in the case when nearly 100 million egg-laying chickens have had to be destroyed recently. People will make their own decisions regarding the meat–vegetable balance, but it has been pointed out that large animal farming provides total dietary protein but may not be SE-optimal when viewed from 40,000 ft. Finally, poor agricultural methods contributed to desertification and perhaps even downfall of ancient civilizations. The dust bowl of the 1930s here almost did us in. A federal government program in the 1930s planted over 220 million trees to stop blowing topsoil that was devastating the great plains.

11.13.2 Construction

Back again to the USAF experience. In 1999 timeframe, the AF used cinder blocks from demolished buildings to successfully contribute to the formation of artificial reefs off the Gulf Coast. Needless to say, metal plumbing fixtures and pipe as well as wood infrastructure all have value both in recycling as well as MSW avoidance. Concrete can be crushed and reused as aggregate to mix with Portland cement, and yes it does pay to do this. Construction demolition waste is rich in metals, plastics, and glass, and is cheaper to recycle than to dispose of in MSW landfills. Just run the numbers!

Cement Technology Lab, the research arm of the Portland Cement Foundation, reports that recycling of old concrete as aggregate can displace as much as 30% new material avoiding mining operations. A definite SE design option winner [7]!

11.13.3 Electronics

Electronics of all kinds, televisions, computers, cellphones, VCR, and DVD players get tossed on a regular basis. There is no reason that any of this should find its way into landfills, or for that matter why it should not all get recycled in one way or another. The use of pyrolysis and gasification can separate organics from inorganics, e.g., glass and metals while safely generating electricity or heat.

Manufacturing of electronics is also a focus of SE—exporting such waste should be prohibited as most offshore recyclers are in gross violation of USEPA and OSHA laws.

11.13.4 Metals—Mining the Nations Landfills

An enormous quantity of metal ends up in landfills every year due to inadequate or inept recycling all across the nation. As can be imagined, keeping this material out of landfills should be job one. As such, the landfills themselves represent a large mining opportunity. It is far less expensive to mine MSW sites than tearing up mountains and pumping acids in to leach out ore. Once done, the mined MSW sites can be reclaimed for reuse. The land values alone, which are rarely considered, will make this an SE cost-savings winner.

In fact, I view this opportunity much the way the electric power industry views government tax credits for purchasing hybrid cars and high efficiency HVAC two large energy consumers in most of our lives. The power generation folks figured it out long ago … It is cheaper to pay people to be more efficient with electricity than it is to build new electric generation facilities. Of course, this is helped by the existence of a quasi-regulated electric power monopoly.

Aluminum, copper, iron, zinc, lead, and mercury are just a few metals of importance. Unlike classic mineral mining that leaves land scarred, the completed MSW "mines" can be recovered for community land use or business development. This is true sustainable development that can continue on an ongoing basis to process real-time waste. The landfill of the future should be a temporary holding, or "surge tank" in the ground, with MSW coming in at one end, and clean energy and recovered materials at the other end. Yes, it will make money. Just run the numbers, but do not forget to include the enormous landfill "forever" closure costs that occur now with little material recovery. Traditional mining will also lead the way in discovering and developing new battery materials.

11.13.5 Paper and Pulp Industry

Forestry in the USA, by definition, must adhere to sustainable development. A tree harvested must be met with a tree replanted. The paper industry is one of the earliest users of recycled materials. Waste disposal at paper mills has also been greatly minimized over the last 25 years. Paper, cardboard, and wood from construction sites are all recyclable.

Example—even plastic is being mixed into wood along with other post consumables to make a plastic wood material suitable for outdoor use in playgrounds, decks, etc., greatly extending or eliminating the use of new wood in those application. Note: these products must meet or exceed construction standards. The Trex company products, for instance, available for about 20 years now are not used directly as structural members, but have other large uses that offset wood felling.

Law of unintended consequences (LOUC)—Note: Recent incident regarding the introduction of Chinese manufactured wallboard [8] has caused a potential $500 million dollar boondoggle as the Chinese manufacturers' idea of recycling seems to have included placing their waste products into our walls being used for new and remodeling-based construction. Be on guard for cheap, imported products coming in with no inspection criteria from countries having no standards whatsoever. That especially includes food. Do not assume the US importers will check this. In the case of the wallboard, they did not.

11.13.6 Plastics and Polymers

A very large industry with products derived largely from petrochemicals. These can be made from agriculture products as well. Potentially future use of crude oil should focus on this area as other renewable fuels come on line for vehicle use. 3D printing will demand ever more sophisticated materials and may itself contribute greatly to SE if material waste can be avoided by printing only what is needed. Gasification that creates H_2 and CO can be converted into plastics as well alongside integrated power production. This will also save energy, the perfect SE nexus. New, less toxic materials (think PFAS) will also come up in this area.

11.13.7 Summary

By the very nature of the depth and breadth of engineering, this has been cursory at best. I included it to demonstrate to all my engineering colleagues a jump-started dialog on SE in their respective disciplines and industry segments. It also serves to remind my ChE colleagues of the importance of a close interaction with all the disciplines to assure an optimal design and finished product.

References

1. Composting, Composting. Dairynet.com
2. Razzaz L Coffee ground composting. www.AustinGround2Ground.org
3. Concrete Recovery. wbcsdcement.org and ashgrove.com
4. Photovoltaics, Nuclear Engineering, Geothermal, National Academies of Science and Engineering. nationalacademies.org
5. http://www.asphaltroofing.org/asphalt-shingle-recycling-faqs, http://asphaltshinglegrinding.com
6. Central District Heating, District Heating. districtenergy.org
7. Cement Technology Lab (CTL). www.ctlgroup.com
8. Chinese Wallboard Toxicity. http://www.atsdr.cdc.gov/drywall/

Chapter 12
Sustainability Engineering Design Resolution Roadmap: Where Do We Go from Here?

Sustainability engineering represents a way forward, above and continuous improvement beyond the existing process and product design engineering methodology. We have covered a lot of ground thus far. The basic roadmap is gleaned from the preceding chapters, with highlights summarized below. There are a few nontechnical programmatic items to help complete the path forward along the way toward normalizing SE.

The key technical improvements over existing classical process design that SE design focus brings to bear include:

12.1 SE-Improved Process and Product Design and Engineering Module

1. The use of *renewable material and energy resources*
2. Incorporation and use of all MSW and other organic waste as feedstocks or energy
3. Integrating power and chemical production to maximize resource conservation amongst disparate processes
4. Focus on equipment and process *efficiency improvements* to avoid suboptimal SE design
5. Development of new SE standards to accommodate closer coordination amongst tightly coupled process and project elements as demanded by SE design practice
6. Deeper, upfront ESOH review to ensure minimized effect on the environment, health, and safety
7. Quality management to assure broader assimilation of all SE designs

But wait, there is more ….

J. P. Perl, *Sustainability Engineering for Enhanced Process Design and Manufacturing Profitability*, https://doi.org/10.1007/978-3-031-52363-2_12

12.2 There Is More Left to Do to Cement SE Design in Place: Some Nontechnical Essentials

Structure is important to make sure SE survives management change, just like OSHA MOC!

12.3 Revitalized US Manufacturing + Industry and Academic Standards

As SE design-based manufacturing becomes more efficient and self-sustaining, it creates new business lines. Greater efficiency will lead to an increase in domestic manufacturing operation opportunities as things become cheaper to make and manufacturing more automated. These are results largely of technical-a and process-related improvements.

All quality management programs (Chap. 9) require program-specific training in order to succeed. Again, we cannot know what we do not know. This is at least partly why we go to college, and it is also why we need to train the engineering and plant operations–related workforce as well, so they are properly prepared to contribute. The former is happening, but needs to be tuned up, and we will get to that later, while the latter has fallen by the wayside, sort of like an orphaned baby that bounced off the last wagon in a wagon train and there was not anyone watching out when it happened!

Here are the three additional programmatic, nontechnical, but essential SE element areas in need of continuous improvement that involve training and education to support the above, namely:

1. A return to the preparation of a technically competent manufacturing workforce
2. Focus on minimum industry practice competency standards and
3. Fine-tune baccalaureate academic preparation, harmonized with industry standards and needs

12.3.1 Technically Competent Work Force: Skills and Trades Necessary for a Sustainable Industry and Economy

The shop classes I took in high school have all but disappeared, leaving a huge void in the preparation of a skilled trade workforce capable of contributing to the new SE-based industries coming into existence. That preparation served to inform my engineering practice and also led to well-paying jobs for many of my fellow HS graduates who chose that route. This mistake needs to be corrected and I have a bit more on this in Appendix A. Remember the importance of “Asking Questions” and “Listening to the Workforce for Input,” the hallmark of successful teams.

12.3.2 Minimum Industry Practice Competency Standards: What Practicing Engineers Need to Know

The NCEES creates and maintains the Professional Engineering (PE) Licensing exam (Chap. 2, Sect. 2.15) for all disciplines including surveying. Having been involved over the last 14 years on that PE exam committee has led to the realization that not only does this establish a minimum practice competency as intended, but the exam specification itself also represents the practice, and therefore requisite preparation curriculum for undergraduate chemical engineering. More details on this can be found in Appendix B.

12.3.3 Baccalaureate Academic Preparation in Harmony with Industry Standards: The Engineer in Training

I used the PE exam specification to inform my curriculum development used in the 6 years I taught the senior chemical engineering capstone design course at the University of Illinois at Chicago (UIC). I only taught the two-semester course, but experience with a fall pre-exam I gave all my seniors suggested the need to more closely harmonize industrial baccalaureate educational standards with academia. As we already have the finest engineering education on the planet, this is only a tune-up, not a revamp and as such should be considered as a continuous improvement process. This will require an industry–academia interaction of the type I enjoyed during my tenure at UIC, and will lead to great benefits for both institutions going forward. Appendix C presents my university ChE design course curriculum that utilized those standards. Much credit and thanks are also due to the dozen or so fellow Chicago area ChE design engineers who helped create the annual spring student design problem as well as serving as mentors to the entire senior class.

12.4 SE Workforce and Revitalizing US Manufacturing: A Cursory Look at Some Low-Hanging Fruit

The explosion of interest over improving efficiency of domestic energy consumption via USEPA and DOE Energy Star programs and chronicled by the EIA and AEE has unexpectedly opened other opportunities regarding a previously weak topic, namely US manufacturing. The cost of manufacturing is largely made up of energy and raw materials, with labor to a lesser extent than ever before owing to an ever-increasing level of automation. This continues to increase quality, and the auto industry is one example of how we can still make things in the USA competitively.

However, an adequately trained and sized workforce is required to move forward. SE-educated engineers and skilled tradesmen will be needed. And the latter need is no different than the spike in trained IT staffing at the associate level that arose in response to industry needs over the past two decades. These good IT people made/make it possible for companies to support the impressive computer capabilities that are the very basis of modern US business. The same is true of skilled SE labor of the future and present and the ramp-up needs to start now.

12.5 A Few Manufacturing Example Potentials That Will Arise from an SE-Design Focus

12.5.1 Natural Gas Fracking

Over the past 10+ years fracturing-enhanced natural gas and oil production has greatly increased US energy production, while lowering NG price. (Note: There continue to be energy-pricing fluctuations, particularly as a result of the ongoing war between Russia and Ukraine.) As we have discussed earlier, natural gas can be gasified and turned into a number of highly value-added merchant products. This is placing unexpected positive demand on domestic US manufacturing to increase, in contradiction to all who were saying our manufacturing was dying and nothing could be done to change that … much like the famous report issued by the US Department of Interior in 1920 timeframe that said we were soon running out of oil [1].

12.5.2 Municipal Solid Waste Recovery

In addition to new resources from fracking, a veritable cornucopia of reusable material is waiting to be mined from our nation's MSW landfills, aka garbage dumps! Better still, consider diversion of such nonhazardous waste direct to energy and materials recovery facilities, many not yet planned let alone built. When viewed as a profit stream for your local plant, MSW may not be large, but the vast amount stored already around the country, well that is another story.

Note: This may require further regulatory development for permitting requirements.

12.5.3 Manufactured Gas Processing Sites

Contamination from old manufactured gas waste sites can also be gasified to avoid off-site treatment that is still being done. These activities require minor adjustments in law, but major realignments between corporations, EPA and well-intentioned community activist groups that sometimes work at cross purposes to sustainability. By the way, some of these regulatory "adjustments" center more on interpretation of existing law that always promoted pollution prevention.

✈Hint—Draw the appropriate quality circle, assemble the quality team, and get it done!✈

12.5.4 Ocean Water Desalination

The process of removing salt from water is either energy intensive or requires expensive, specialized filtration. Our coastal regions, e.g., California, are highly focused on agriculture. These regions have been drawing down on precious groundwater causing ground subsidence with attendant structural damage, and lack of potable drinking water. Application of SE design might include the co-location of desalination by evaporation, with integrated energy production and chemicals, e.g., fertilizers. This is the ultimate level of SE. Once governments factor in the true cost of supplying drinking and agriculture water, these can be factored into competent, fact-based SE designs.

12.5.5 Generalized Integrated Power and Chemical Production

There are innumerable combinations of exothermic and endothermic reactions in the chemical manufacturing business, constantly heating and cooling various streams. In addition, the material exiting process A, per USEPA RCRA (Chap. 6, Sect. 6.5.7) , is not a waste until you say it is, and certainly not if it can be used as a feed to process B. Matching wastes and feeds is not exactly new, but teaming disparate processes along with power production is still in its infancy. Finding and matching these is both the SE vision and challenge.

Appendix C has two examples of fully integrated power and disparate chemical production. These were created and used by my colleagues and myself in the two-semester capstone course I taught at UIC. I am still searching for actual examples, but at least some companies are starting to generate their own power and some are combining disparate processes for both energy and material optimization.

12.6 SE Bridges: Getting from Here to Tomorrow

Some of the above examples must be taken as a sort of quasi-SE approach, as fracking, e.g., merely extends a nonrenewable hydrocarbon resource, but it does, however, free up petroleum to greatly extend use as a transportation energy source. Similarly, a 54-mpg mandate doubles the fleet average vehicle performance, but is still based on a nonrenewable resource. These elements can buy us time to get into economically competitive photovoltaic, algae-to-oil, NH_3, H_2, and other as yet undeveloped renewable materials and energy sources. This is the point of establishing an SE index.

We cannot just grow our way out of this, but agriculture does have a role here. Some say feeding cattle corn, etc. is not the most efficient way to obtain human nutrition. A proper SE analysis of this would be most useful as we move away from a total reliance on meat over to a more balanced one including more grains with fiber content. Remember to keep this apolitical. Just run the numbers. (Include subsidies!)

As high-value-added manufactured products, such as computer chips and automobiles, become so highly automated, it makes less sense to pay for importation from offshore production facilities. As usual, there are advantages to production of items for sale in the host country, not so much to be transported back as imports. Again—Just run the numbers and look for energy generation opportunities. Of course, balancing and maintaining healthy trade relations is key.

12.7 Sustainability-Driven Industry Recovery and Rebuilding

Here is a drill-down look at one of these examples, MSW, a huge arena ripe for SE manufacturing.

12.7.1 Landfill Municipal Solid Waste (MSW): New Integrated Energy and Chemical Production Opportunity

According to the EPA, "in 2013, Americans generated about 254 million tons of trash and recycled and composted about 87 million tons of this material, equivalent to a 34.3 percent recycling rate. On average, we recycled and composted 1.51 pounds of our individual waste generation of 4.40 pounds per person per day."

The USA generates an enormous quantity of MSW and has done so in a large way for the better part of a century now. MSW includes all post-manufacturing by-products that are not considered EPA-regulated toxic or hazardous materials, including glass, plastic, and metal values not recycled. In addition, an enormous quantity of household MSW (garbage) is also generated and is available for resource recovery. Another category, special waste, enjoys an intermediary regulatory standard. Some facilities are setup to burn MS as "refuse derived fuel" (RDF). Many landfills have collection systems to recover and utilize natural gas that evolves from natural anaerobic digestion of waste organic materials, but this still leaves valuable material and the constant threat of land subsidence. ✈Could make the Golf Courses more interesting as they change annually!✈

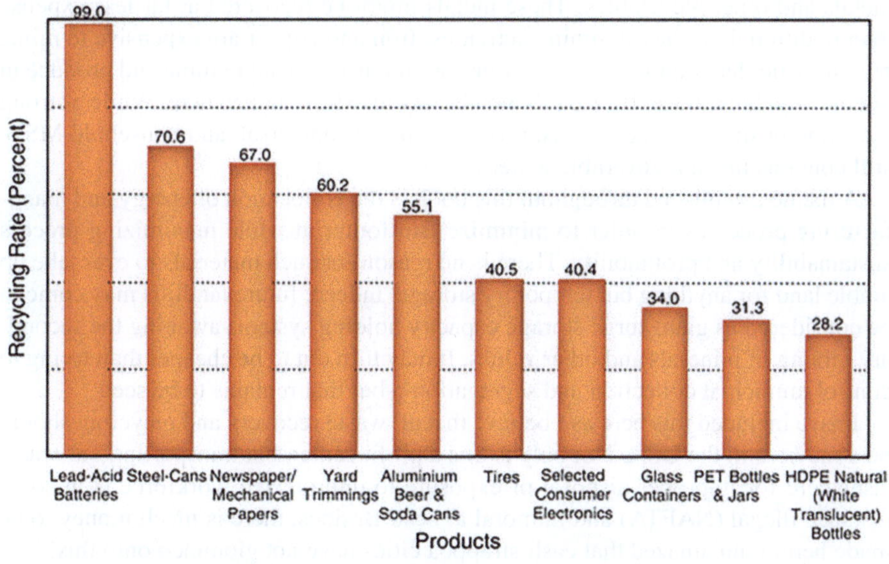

Fig. 12.1 MSW recycling rate of various products (from USEPA)

It is obvious that there is an enormous business opportunity to capture resources that are not recycled. As mentioned earlier, simply capturing landfill methane is "necessary but not sufficient" for SE. Collocating power and manufacturing, particularly on large MSW sites, can provide the necessary economics of scale, and the flow of feedstock, aka garbage, is nearly continuous!

Figure 12.1, from the EPA website [2], gives some idea of the potential business opportunity for recycle improvement.

12.7.2 The End of the Landfill as We Know It

Any consideration of sustainability engineering must take into account the disposition of all MSW as many industries generate a fairly large quantity of MSW. This includes paper, food, plastics, glass containers, and metals of all kinds. Always, the Btu is the coin of the realm and care must be taken to assure a net positive gain in Btu whenever recycling or reuse options are examined. Disposal of MSW into landfills should consider the ultimate cost of reclaiming these diluted values as this will surely need to be done in the not too distant (20–40 years) future as we run out of otherwise arable land. In fact, these lands should, in the future, be designed more as temporary, or, in CPI terms, surge storage facilities much as we do with tanks in chemical manufacturing.

This also brings to mind the need to consider as an industry itself, mining of existing landfills, perhaps at the end of bio-energy recovery, to obtain valuable metals and other recyclables. These metals might be recovered at far less expense than traditional ore-based mining activities. Iron and copper are expensive to mine, but, in particular, aluminum is a very energy-intensive metal to mine and produce in quality suitable for use. Rare earth metals may also be obtained here. While a strong recycling business already exists, it is by no means universal, and household MSW still contains much recoverable material.

A theme established throughout this book is the collocation of energy and manufacturing processes in order to minimize Btu footprint while maximizing process sustainability and profitability. There is no reason for such materials to ever take up arable land for anything but temporary storage. Indeed, future landfills may come to be considered as giant surge storage capacity holding systems awaiting the secondary mining of minerals and other values. It may turn out to be cheaper than trying to control municipal collection and segregation—but that remains to be seen.

I have included this here as I believe that all waste recovery and recycling should be done *here* in the USA. Not only is this a profit center, but transporting our waste elsewhere for improper disposal or exposure to unprotected workforce is in some instance illegal (NAFTA) and immoral at best. Besides, there is much money to be made here. I am amazed that cash-strapped cities have not glommed onto this!

As mentioned throughout this book, state, local, and federal governments have collected and continue to generate quite a library of useful data and information regarding energy and materials disposition in the USA. I have put samplings of useful ones throughout the book. Here is another example:

12.7.3 Energy Recovery from Waste (Excerpted from USEPA) [3]

Energy recovery from waste is the conversion of non-recyclable waste materials into useable heat, electricity, or fuel through a variety of processes, including combustion, gasification, pyrolysis, anaerobic digestion, and landfill gas (LFG) recovery. This process is often called waste-to-energy (WTE).

12.7.3.1 Energy Recovery from Waste Facility

Energy recovery from waste is part of the non-hazardous waste management hierarchy. Converting non-recyclable waste materials into electricity and heat generates a renewable energy source and reduces carbon emissions by offsetting the need for energy from fossil sources and reduces methane generation from landfills.

Defined as separated yard waste or food waste, including recycled cooking and trap grease, and materials described in §80.1426(f)(5)(i). Final regulations allow separated municipal solid waste (after all recyclable materials have been removed) to qualify as "separated yard or food waste."

12.7.3.2 Combustion or Reformatting of MSW with Energy Recovery

Currently there are 86 facilities in the United States for combustion of municipal solid waste (MSW), with energy recovery. These facilities are located in 25 states, mainly in the Northeast. No new plants have been built in the US since 1995, but some plants have expanded to handle additional waste and create more energy. The 86 facilities have the capacity to produce 2,720 megawatts of power per year by processing more than 28 million tons of waste per year. According to Municipal Solid Waste in the US: Facts and Figures, in 2011 we combusted about 29 million tons of MSW (about 12 percent) for energy recovery. After energy is recovered, approximately ten percent of the volume remains as ash. This ash is generally sent to a landfill. Please visit EPA's Landfill Methane Outreach Program for additional information on how energy is recovered from landfills.

→Landfill ash may be used as aggregate in cement as well as wood plastic, or possibly in ore smelting to recover metal values—so really nothing should be left behind!→

12.8 Improving US Economic Performance, Safety, and Profitability Through Sustainability Engineering

As already noted, companies are recognizing the value/savings of producing onsite electricity. Some are even moving to go off the grid, but this continues to be a challenge. As more manufacturing moves toward integrated energy and product production, *sustainability engineering* will lead the way into a more secure, safe, and profitable future for the USA. We have already proven that human productivity continues to increase. The same is now true in material and energy resource husbandry as the focus on sustainability continues to show profitability just as pollution prevention does now and as it did in the past. The mere vastness of the USA bodes well for richly rewarding sustainability efforts.

References

1. Olien DD, Olien RM (1993) Running out of oil: discourse and public policy, 1909–1929. Bus Econ Hist 22(2): Winter. www.thebhc.org
2. Municipal Solid Waste (MSW) Materials recovery from MSW. http://www3.epa.gov/epawaste/nonhaz/municipal/
3. Energy Recovery from Waste. WastetoEnergy https://archive.epa.gov/epawaste/nonhaz/municipal/web/html/index-11.html

Appendix A
SE and Revitalized US Manufacturing

"You're the first person who ever asked for our opinion," I was told after asking a shop floor worker for his input while troubleshooting at a major Midwest manufacturing plant.

I will never forget that conversation and how useful it was going forward. The good people who operate the manufacturing facilities that engineers design, build, and manage are their ears and eyes, "boots on the ground" element that keeps things going. Good engineers understand this as well as the need for them to be there to assist in interpretation and rolling their floor observations into useful improvements. The technical workforce needs a seat at the quality table, or risk suboptimal operation.

A1.1 An SE-Prepared Manufacturing Workforce

As described in Chap. 9, all quality management programs require program-specific training. Again, we cannot know what we do not know. This is at least partly why we go to college, and it is also why we need to train the engineering-related workforce as well, so they are properly prepared to contribute. The former is happening, but needs to be tuned up (Appendix C), while the latter has fallen by the wayside, sort of like an orphaned baby that bounced off the last wagon in a wagon train and there was not anyone watching out when it happened.

In researching this workforce topic I came upon the Hollings Manufacturing Extension Partnership Program, under the National Institute of Standards and Technology (NIST), formerly the National Bureau of Standards (NBS) [1].

I suggest you visit this site to learn about the present US shortfall of engineers and skilled technicians needed to fill well-paying jobs.

As high value-added manufactured products, such as computer chips and automobiles become so highly automated, it makes less sense to pay for importation

J. P. Perl, *Sustainability Engineering for Enhanced Process Design and Manufacturing Profitability*, https://doi.org/10.1007/978-3-031-52363-2

from offshore production facilities. As usual, there are advantages to the production of items for sale in the host country, not so much to be transported back as imports. Again—Just run the numbers.

When I graduated Lane Technical High School in Chicago in 1970, I had taken every conceivable shop class I could. Machining, foundry, electrical, electronics, and woodworking were my choices, but automotive, aviation, and printshop, where our daily newspaper and annual yearbook were published rounded out the offerings. In addition, most of the shop classes had second-level courses that served as technical career path entryways for non-college bound students. These graduates were actually recruited right out of our high school, into good paying jobs. Vocational training is making a comeback somewhat in junior and community colleges, but should be present in high school as well, just as S.T.E.M. is finally doing the same. In fact, counselling, and more specifically, mentoring of all types should start in middle school.

Fast forward to the present, and these high school shops have all but disappeared, a self-fulfilling prophecy of the domestic manufacturing doomsters. Government educational budget shortfalls and the misguided notion that everyone needs a 4-year college degree have also contributed to this unusual predicament. There is no pipeline of trained machinists, tool, and die makers, etc.—so offshore it goes, no matter how automated the factory. IBM, prior to selling their PC manufacturing business to Lenovo actually created a totally automated PC factory in New York State … so why did this leave?

But here too, it is not clear without analysis what path to take. Having domestic technical capabilities in these areas, however, assures our competitive edge into the future. Who would have predicted major cities creating bike lanes on our crowded streets, or a return to greater reliance on public mass transportation to get from point A to point B, but that is what is happening. We are returning to exercise and this will lead to better health outcomes and lower medical bills, etc. This is a prime example of SE thought process. I believe most people sense this connection—hence the huge public interest in "All Things Green."

A1.2 US Economic Sustainability

In order for the US Economy itself to be sustainable, we will have to recover some of those old programs that worked. Perhaps an old-fashioned European Trade Guild program like the one that my Father attended over 3 years in the late 1930s and early 1940s Budapest, Hungary. After completing the eighth grade, Dad entered trade school and then became a master tailor, through training and apprenticeship. This preparation ultimately allowed him to provide quite well for his family of five after immigration to and marriage in the USA following the war. There is great need for trained professionals throughout the USA. I am amazed and grateful for the number of immigrants who come here and do the jobs that Americans have been told do nOT exist anymore!

On September 2015, President Obama closed the offshore profits tax loophole in an effort to bring that money back along with manufacturing. This might bode well for sustainable manufacturing, if we can get enough trained technicians in place! Shortfall of engineers suggests opening the door to visa's for the educated seeking to migrate to the USA while US Universities increase enrollments.

The graphic on the following page is from the National Institute for Standards and Technology (NIST) website nist.gov [1] and describe interesting facts regarding the state of manufacturing in the USA today. As already discussed in this book, this graphic identifies a missing link necessary for SE success, that is, a technically qualified workforce. Let's bring back the old European trade guild system.

Excerpt from NIST on US Manufacturing:

At a time of opportunity and challenge for U.S. manufacturing, NIST is working with industry and universities to develop essential measurement capabilities and to forge precompetitive collaborations that help U.S. manufacturers overcome shared technical obstacles. A partner to the U.S. manufacturing sector for more than a century, NIST has a proven track record in delivering useful tools and technical assistance that existing manufacturers and aspiring start-ups need. Timely technical assists from NIST can help the nation's manufacturers to invent, innovate and create new products and services more rapidly and more efficiently than their competitors around the world. (https://www.nist.gov/manufacturing)

→- Instead of drugs for the so-called attention deficit, how about giving creative, talented people who do not fit the college mold something extremely useful to do with their hands!-JPP→-

Reference

1. National Institute of Standards and Technology (NIST). Need for trade workers. www.nist.gov - https://www.nist.gov/system/files/making-it-in-america-v11.pdf

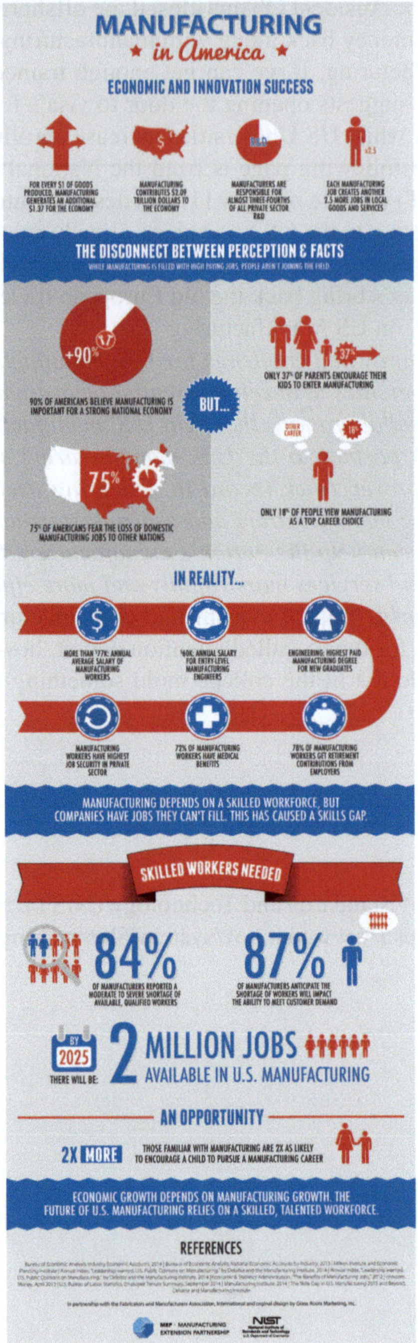

Fig. A1 Skilled worker shortfall, both training and people

Appendix B
Chemical Engineering Discipline, Practice and Professional Standards (NCEES Standards and a Practice-Based Senior Design Course)

B1.1 Introduction

When I started teaching design in 2008, I was asked to create a ChE 101 lecture for presentation to entering freshmen in the College of Engineering, University of Illinois at Chicago (UIC) to give entering engineering students an idea of what their chosen major was, and a last chance to switch! What follows is a really brief and not perfect review of chemical engineering. Many will take issue perhaps of where, when, and how ChE arose. Much of this I learned while in graduate school, mostly on my own, so the errors are all mine. Prof. Google also provided much, particularly regarding the practice in England, which appears to have emerged around the same time as in the USA, in the late 1800s. Several excellent, detailed compilations can be found in the references provided [1–3]. Almost immediately, professional engineering licensure [4] began to emerge in various states as a matter of public safety, to prevent boiler explosions and bridge collapses, to name just a few. It is of interest to note that other public safety regulations, such as the Pure Food and Drug Act arose around the same time and for identical reasons.

B1.2 Chemical Engineering Discipline and Preparation

The combination of practice and academia of chemical engineering had its early start at MIT in the USA with Arthur D. Little, ca 1888 and "Unit Operations" and in the UK with George Davis, at the University of Manchester, ca 1887. In the USA, the first 4-year curriculum was developed at MIT in the late 1880s and was a clear, purposeful, and successful attempt to translate Little's organized approach to industrial scale chemical production into a organized, teachable academic curriculum [2].

J. P. Perl, *Sustainability Engineering for Enhanced Process Design and Manufacturing Profitability*, https://doi.org/10.1007/978-3-031-52363-2

Little and colleagues at MIT developed the "Unit Operations" approach still practiced today. This approach was extremely successful in that it created a scientific basis for engineering principles and practices developed by Little over his more than 25 years of practice. The original MIT ChE faculty was largely scientists and welcomed Little and his industrial bent and interest. Indeed, the science-based American Chemical Society (ACS) Industrial and Engineering Chemistry Division (I&EC) almost became what eventually turned into AIChE in 1908 [5]. For many years, ChE and Chemistry faculty alike continued to publish engineering works and developments in the ACS I&EC Journal.

As developments progressed, ChE Unit Operations developed mathematical methods such as the McCabe–Thiele approach for fractional distillation design, as well as others that were major advances in the science of ChE. Early rigorous scientific methods in ChE R&D labs in academia as well as industry bore enormous fruits in vapor–liquid, liquid–liquid, and solid–liquid equilibria needed to advance engineering technology and design beyond the simple know-how, allowing organized process and product design optimization to emerge as a permanent process design methodology. Most of this was aimed at production improvements and featured empirical studies and resultant correlations that greatly improved our ability to design from the book.

To me, engineering is know-how while science is know-why. Engineers are chiefly interested in how to do things—the why can come later as in the Edison approach. While engineering is not applied science, as I often hear, it still depends on the firm use of, and the essence of the scientific method, i.e., elucidation of physical descriptions, or, mechanisms to explain observations in industry or the world at large—repeatable anywhere in the world. Thomas Edison, "The Wizard of Menlo Park," understood this and created the modern R&D lab approach to combine the best of basic, trial, and error observation, with solid physics, chemistry, and math [6].

The publication of "Transport Phenomenon" by Bird, Stewart, and Lightfoot (BS&L) in 1960 (Chap. 2, Sect. 2.5) set the ChE discipline on a more rigorous but somewhat industrially disconnected approach to ChE. The theories of electromagnetism of James Clerk Maxwell [7] a century earlier used a similar mathematical approach to explain Faraday's observations on electromagnetism [8] and indeed led to electromagnetic transport eqs. [9], and the modern field of electrical engineering. BS&L, of course, with the generalized continuity equation evolved Transport Phenomena (Chap. 2, Sect. 2.5).

After WWII in the USA, the government realized the importance of sponsored research that industry, left to its own devices, would never undertake. The then nascent field of Transport Phenomena was ripe to support this and set ChE academia on a golden age of research. Similarly a renaissance of physical chemistry in chemical engineering occurred to better understand molecular interaction beyond the ideal gas laws and van der Waals, virial, and other equations of state. As a result, it is often difficult to distinguish an academic chemistry from a chemical engineering research lab. The completion of my PhD in chemical engineering at IIT was followed by postdoctoral research at Brown University, Department of Chemistry,

NSF Materials Research Center! (Quite an honor for this Engineer) I then emerged back into more traditional engineering, and this scientific preparation helped me elucidate the need for appropriate matching of the two diverse approaches.

B1.3 The NCEES Practice-Based National Licensing Exam for Professional Engineering

The BS ChE is considered a terminal degree in that it is possible to practice successfully with no additional formal education. PE licensure, however, requires continuing education of a professional nature to keep up with advances. Although some 80–90% of ChE students go on to industrial careers, they are increasingly educated by faculty whose sole experience is academic, science-based research.

The industrial practice of ChE, however, requires competency in a number of core practice areas as outlined by the NCEES [10]. For the last 15 years, I have worked on the PE ChE Exam Development Committee. This volunteer member group meets twice a year to review what it deems to be the minimal competency for the licensed practice of ChE. This group has been meeting for many decades now and includes members from industry, government and academia. The professional engineering exam is assembled by these men and women, who volunteer their time at the NCEES Greenville, SC headquarters. As mentioned previously, I developed a teaching curriculum that incorporates the PE Exam Specification as well as my own professional experience to guide the senior capstone design course I taught at UIC for 6 years starting in 2008. This was augmented each spring by six to ten industrial colleagues serving as student design team mentors, who also helped create a novel, practice-based student design problem for the second semester of each year. This is described in greater detail in Appendix C.

B1.4 The Case for Engineering Graduate School

As noted, US chemical engineering got its start at MIT in the late 1880s, but they did not commence granting PhD degrees until the late 1920s [1]. They felt that this was unnecessary and superfluous for industrial practice preparation and their MS was practice oriented. Modern US MS level course work, however, is largely transport phenomenon–based and certainly has found application along with molecular thermodynamics, in the field of process computer modeling. Aspen Plus, Hysis, and ChemCad are just a few modeling programs that have benefited greatly from enhanced research in the area of thermodynamics, particularly in the area of molecular thermodynamics as well as kinetics. Transport-based physical process modeling of any kind continues to inform the operation and control of complex interactive refineries and other similar enterprises. Recently, schools like Penn have started to emphasize new product as well as process design, clearly a natural follow-on to academic research. Think of developments in nanotechnology, for example.

It is dangerous and unfair to generalize the relative importance of undergraduate and graduate education. Clearly obtaining a theory-based MSChE can help inform the practitioner as far as understanding the underlying math and physics of the field. It is not only the province of R&D to come up with fundamental advances, but an MS can help one think more like a chemical engineer, coupling practice-based experience with theoretical models. The PhD level can help make large leaps in theory that only multiple years of research and focus can bring to bear. Some form of work/study might help here, e.g., the MIT Practice School approach. But it all gets back to practical, safe, integrated, sustainable chemical manufacturing and energy generation processes.

Without academic research there can be little major advance; without industrially relevant academic preparation, however, our engineers risk falling behind in a very competitive world. Some schools have started a focus on product design in addition to process design, and it is hoped sustainability will be their guiding light.

So I found in the NCEES professional engineering licensure process, my practice, and adjunct university design teaching, a way to reconnect academia with industry. Perhaps this will provide a feedback loop for continuous improvement and advance of the field. Incorporation of SE design principles into undergraduate teaching is a needed step in the quality continuous improvement process.

References

1. Chemical Engineering at MIT, The Improbable Achievement ChE at MIT, 1979 by MIT Dept. of ChE
2. History of Chemical Engineering at MIT. http://web.mit.edu/cheme/about/history.html
3. Phillip CW (2008) 200 years of ChE pedagogy. AIChE 100th Anniv., AIChE Annual Mtg., Philadelphia
4. NCEES. NCEES.org website has all pertinent information on Engineering Licensing in the US
5. American Institute of Chemical Engineering, AIChE. www.aiche.org
6. Thomas Edison. http://www.edisonmuckers.org
7. Maxwell explaining Faraday's electromagnetism. http://www.maxwells-equations.com/faraday/faradays-law.php
8. Faraday M. The chemical history of a candle. Lecture series, Royal Society of London, ca 1860
9. Perl JP. PhD Thesis, 1984, Complex microwave dielectric properties of liquids, solutions, and emulsions, May 1984, Illinois Institute of Technology
10. NCEES Exam Committee. Professional Engineering Exam Spec.—Chemical, NCEES, 2011, Clemson, SC
11. Perl J, Brown S, Palmer J, Keesom W, O'Brien D, Shannon P, Rusinak D (2011) A PE based industry-academia cooperative chemical engineering design course. Perl to NCEES, May 2011

Additional Reference

- Finch JK (1960) The story of engineering, Doubleday anchor

B2.1 ChE NCEES Professional Engineering (PE) Examination Specification (A Model Educational Standard) By Jeffery P Perl, PhD, PE

B2.1.1 Introduction

The history of engineering registration dates back to Wyoming law in 1907, and by 1950, all states plus Alaska, Hawaii, the District of Columbia, and Puerto Rico had engineering registration laws (NCEES website). The overarching purpose is the protection of the public by establishing minimum competency requirements for the practice of engineering [1].

The National Council of Examiners of Engineers and Surveyors (NCEES) create and maintain professional practice standards and licensure examinations in all member disciplines. I have served as a member of the committee that produces and watches over the PE licensure of chemical engineers for going on 15 years now. As such, I became very interested in the existing examination specification as it serves to guide and inform the creation of practice-relevant PE and fundamentals of engineering (FE) examinations. These specifications are openly available at https://ncees.org/ website. The committee consists of industry, academia, and government members. The committee watches over and vets additions to the examination as well as monitors its performance along with a professional examination metric company, Pearson Vue.

Not long after joining the NCEES committee, I was invited to join the faculty as Adjunct Professor of Chemical Engineering at (UIC). I was hired, as a PhD practitioner, to teach the two-semester senior process design course (Design I and II). This is the ChE capstone course where UIC students take what they have learned over their first 3 years of study, to create, or, design a chemical production process. I created a syllabus for both semesters out of whole cloth based on my professional work practice as well as preparation by one of my graduate school mentors, Professor Bill Langdon, who invited me to teach the Illinois Institute of Technology design capstone courses along with him while I was still a wee lad in graduate school.

As I prepared my UIC syllabus, I used elements of the NCEES (ChE PE Exam) specification along with my own experience to guide the preparation of budding young chemical engineers in the two-semester course. I found this to be somewhat disjointed from the present academic approach (but not hopeless!) and used the two-semester course to bring to life, with the help of several friends and colleagues, ChE theory as illuminated by practical problem-solving.

The first semester covers basic tools, tips, and techniques necessary to accomplish design as well as prepare for a career in an operating chemical production company. The second semester, I created a design project along with six to ten of my Chicago area design professional colleagues. For each of my 6 years at UIC, a unique student design project was created, all of which introduced the concept of sustainability, the idea being to inculcate sound engineering practices and habits early on for our students long-term career development.

The focus was on integrated power and production as "ripped from chemical engineering headlines."

I was also asked to prepare a ChE 101 lecture on "What is Chemical Engineering?" [2], for the UIC College of Engineering, some of which is summarized at the beginning of this Appendix.

B2.1.2 The NCEES Examination Specification as an Outline for the Practice of Chemical Engineering

The NCEES Chemical Engineering PE Exam Specification can be downloaded from NCEES website [1]. The 2011 version is reproduced here complete with NCEES instructions to students preparing for the PE exam. As soon as I became active in the Committee in 2007, I recognized the Exam specification represented the practice of chemical engineering at least as it was taught to me 40 years ago and as I know it now professionally. As described earlier, feedback from pre-course examinations I administered on day 1 of my university teaching suggested the need for additional review of these subjects as they are practiced, in juxtaposition to theory to better prepare them for their industrial careers. Greater than 80–90% of our students go directly to work as the ChE degree is considered a "terminal degree" in that it is considered sufficient preparation to commence a career in the CPI.

The NCEES ChE subject matter areas and the number of questions used in exam preparation as well as guidance for student study and preparation follows:

> Excerpt from NCEES Website, Courtesy of NCEES (NOTE: More recently the exam has gone from paper to computer. It is administered/proctored by the Pearson VUE Company. Examinees have access only to a digital "Supplied Reference Handbook" created by members of the exam committee.—JPP 2023)
>
> ***NCEES Principles and Practice of Engineering Examination***
> ***CHEMICAL Exam Specifications***
> ***Effective Beginning with the April 2011 Examinations***

- *The exam is an 8-hour open-book exam. It contains 40 multiple-choice questions in the 4-hour morning session, and 40 multiple-choice questions in the 4-hour afternoon session. Examinee works all questions.*

- *The exam uses both the International System of units (SI) and the US Customary System (USCS).*
- *The exam is developed with questions that will require a variety of approaches and methodologies, including design, analysis, and application.*
- *The knowledge areas specified as examples of kinds of knowledge are not exclusive or exhaustive categories.*

Approximate Number of Questions	
I. Mass/Energy Balances and Thermodynamics	***18***
A. Mass balances	*8*
1. Material balances with no reaction (e.g., *phase behavior, mass, volume, density, composition, purge, bypass)*	
2. Material balances with reaction (e.g., *multiple reactions, incomplete reactions, excess reactant, purge, bypass, recycle, combustion)*	
B. Energy balances and thermodynamics	*10*
1. Energy balances on non-reactive systems (e.g., *sensible heat, latent heat, heat of solution)*	
2. Energy balances on reactive systems (e.g., *heat of reaction and combinations with sensible heat; latent heat; heat of solution)*	
3. Power cycles (e.g., *refrigeration, engines, turbines, heat recovery)*	
II. Heat transfer	***13***
A. Mechanisms	*7*
1. Heat transfer without phase change (e.g., *thermal conductivity, heat capacity, conduction, convection, free/forced heat transfer coefficients/correlations, radiation, combinations thereof)*	
2. Heat transfer with phase change (e.g., *vaporization and evaporation, condensation, sublimation, crystallization, latent heat)*	
B. Applications	*6*
1. Heat exchange equipment design (e.g., *overall heat transfer coefficient, fouling factors, LMTD, F-factor, equipment selection, insulation)*	
2. Heat exchange equipment analysis [e.g., *pressure drop, fouling effects, performance evaluation (NTU), changes in parameters]*	
III. Kinetics	***9***
A. Reaction parameters	*5*
1. Rate equation (e.g., *rate constant, activation energy, order of reaction, mechanisms, catalysis)*	
2. Chemical equilibria (e.g., *temperature and pressure dependence, composition)*	
B. Reactors	*4*
1. Conversion in single reactors [e.g., *batch reactor, continuous stirred tank reactor (CSTR), plug flow reactor (PFR)]*	
2. Conversion in complex reactors (e.g., *reactors in series: CSTR and/or PFR, multiphase reactors, fluidized beds, packed beds, recycle, bioreactors)*	
3. Yield and selectivity	
IV. Fluids	***13***
A. Mechanical-energy balance	*10*
1. Flow behavior (e.g., *viscosity; velocity; Reynolds number; friction factor; pressure drop in pipes, valves, and fittings; expansion/contraction; porous media; particle dynamics; fluidization; sonic velocity; laminar/turbulent; two-phase flow)*	

Approximate Number of Questions	
2. Flow applications (e.g., *potential and kinetic energy, friction, flow networks, mixing, pumps, NPSH, turbines, compressors, drivers, solids handling)*	
B. Flow and pressure measurement techniques	*3*
1. Flow measurement application (e.g., *mass and volumetric meters) and pressure measurement application* (e.g., *permanent pressure drop, differential pressure devices)*	
V. Mass transfer	***11***
A. Phase equilibria	*4*
1. Ideal systems (e.g., *Henry's law, Raoult's law, Dalton's law, ideal gas law, vapor pressure)*	
2. Non-ideal systems (e.g., *activity coefficients, fugacity coefficients, azeotropes, immiscible/ partially miscible phases, equations of state)*	
3. Phase equilibrium applications (e.g., *bubble point, dew point, flash, critical states)*	
B. Continuous vapor–liquid contactors	*6*
1. Material and energy balances for trayed units and packed units (e.g., *absorption, stripping, distillation)*	
2. Design parameters for trayed units (e.g., *minimum flow rates and reflux, minimum and theoretical stages, feed location, tray selection, capacity/efficiency, flooding, dumping, tray hydraulics)*	
3. Design parameters for non-trayed units (e.g., *minimum flow rates and reflux, minimum stages, theoretical stages/NTU, feed location, packing selection, capacity/efficiency, flooding, pressure drop, mass transfer coefficients/height of transfer units)*	
C. Miscellaneous mass transfer processes	*1*
1. Continuous, batch, and semicontinuous (e.g., *drying, membranes, extraction, crystallization, filtration, leaching, humidification, diffusion, adsorption, absorption, stripping, distillation)*	
VI. Plant design and operation	***16***
A. Economic Considerations	*1*
1. Cost estimation and project evaluation (e.g., *capital costs, depreciation, operating costs, risk evaluation, optimization, return on investment)*	
B. Design	*8*
1. Process design (e.g., *process flow sheets, P&ID, specifications, procedures, modeling/ simulation, scale-up, process or product development, boundary conditions)*	
2. Process equipment design (e.g., *equipment selection, optimization, design temperature, design pressure)*	
3. Siting considerations (e.g., *security, ingress, egress, plant layout, utilities, natural disasters, human factors)*	
4. Instrumentation and process control (e.g., *sensors, controller actions, feedback/feed-forward actions)*	
5. Materials of construction (e.g., *material properties and selection, corrosion considerations)*	
C. Operation	*3*
1. Process and equipment reliability (e.g., *testing, preventive maintenance, startup/shutdown procedures, robustness)*	
2. Process improvement and troubleshooting (e.g., *debottlenecking, experimental design and evaluation, optimization)*	
D. Safety, health, and environment	*4*
1. Protection systems [e.g., *pressure/vacuum relief valves (safety valves), flares, rupture disks, vents, vacuum breakers, inerting, seal legs, discharge location, configuration, fire protection]*	

Approximate Number of Questions
2. Industrial hygiene (e.g., *MSDS, exposure limits and control, noise control, ventilation, personal protective equipment)*
*3. Hazard identification and management [*e.g., *flammability/ explosive limits, auto-ignition, reactor stability, process hazard analysis, safety integrity level (SIL), management of change]*
4. Environmental considerations (e.g., *emissions evaluation, permitting, pollution prevention, mitigation, waste determination)*

References

1. NCEES.ORG
2. Perl JP History and practice of Chemical Engineering, freshman college of engineering lecture University of Illinois at Chicago, Aug 2008–2014.

B3.1 Sustainability Engineering in Teaching Undergraduate Chemical Engineering Design

What follows is a submittal for an NCEES award for teaching Professional Engineering Design [11]. Although we did not win, it captures the essence of the teaching philosophy that is useful for team building as well as academics. It is also reflective of the practice-based approach already described herein. Tremendous input was had for the second semester of this course from 7 to 10 of my design professional friends and colleagues from Chicago area engineering firms. In this manner, our students all had close personal attention, in the 4 or 5 to 1 ratio. I have placed this section on teaching in this book as it also represents a philosophical approach to Sustainability Engineering that if emphasized early in one's career will provide a lifetime SE approach.

B3.2 "A PE-Based Industry-Academia Cooperative Chemical Engineering Design Course"

Submitted for the 2011
NCEES Engineering Award for Connecting Professional Practice and Education
By
University of Illinois at Chicago, Department of Chemical Engineering
Jeffery P. Perl, PhD, PE, CHMM, Adjunct Professor of Chemical Engineering.
Course Coordinator
Industry Mentors and Student Design Team Instructors

Shannon Brown, PE and Jerry Palmer—Ambitech Engineers
Bill Keesom and Dennis O'Brien, PE—Jacobs Consultancy
Pat Shannon, PE and Dan Rusinak, PE—Middough Engineers
Adam Kanyuh—UOP

Abstract
The capstone chemical engineering (ChE) design course at the University of Illinois at Chicago (UIC) is taught in two semesters. In its present form, the subject of this submittal, the course has been taught over the past 3 years. In addition to meeting ABET academic requirements; it now incorporates relevant elements of the NCEES Chemical Engineering FE and PE exam specifications. The fall semester focus is on traditional design calculations and procedures including economics and environmental health and safety considerations. The spring semester, which is the prime focus of this submittal, turns to eight student-team-based design projects and incorporates outside industry engineers as client/mentors. In all, seven practicing area chemical process design engineers, five whom are, PE-licensed, were called upon to help design the structure of the spring course as well as participate in its delivery, serving essentially as individual team instructors. The course structure is modeled after a chemical industry based conceptual design project scoping and review methodology.

A conceptual level of design is achievable in the second semester and acquaints the students with a focused method of assembling, reviewing and validating a stage-gate approach to total project design evaluation. Spring grades are based primarily on project performance but periodic exams are given based on supporting lecture material, much of which is presented by the industry mentors themselves. Students may opt to take either a final exam or sit for the spring NCEES FE Exam. Over the past 3 years, 57 out of 100 of our UIC senior chemical engineering students have opted for the FE exam. This recent year, 75% (27 students) sat for the April 2011 FE exam. What makes this chemical process design program unique is the industry mentors who spend, with enthusiastic company support, 30 h of intensive student group in-class reviews that simulate those in the process design industry, an additional 30 + h in telecons and student office visits to monitor and enhance project progress, and at least 40 h of preplanning and midcourse assessment and grading during the semester. This design course structure is amenable to all engineering disciplines, i.e., following strict adherence to ABET academic standards/requirements, incorporation of NCEES practice based-discipline specific standards and strong mentoring from working professional engineer design practitioners, to help prepare our young engineers of the future, be they industry or university bound.

1. *Project Description.*

The discipline of chemical engineering concerns itself primarily with the process of converting large quantities of raw materials into finished products, e.g., crude oil into gasoline as well as plastics, food and pharmaceutical production to name a few. This needs to be done economically, safely and, in the face of diminishing natural resources and environmental quality, in a sustainable manner. The UIC senior chemical process design capstone course presents a disciplined, systematic approach to designing modern petroleum refineries and chemical manufacturing facilities to name a few. It does so by drawing from the sum of undergraduate preparation, existing academic and practice standards, tempered by seasoned practitioner-based experience. A more detailed timeline and syllabus is presented in Sect. 6.

What follows describes a new 2-semester design effort that has just completed 3 full years (Fall 2008–Spring 2011) at the University of Illinois at Chicago, Department of Chemical Engineering. In addition to ABET compliance, the senior capstone chemical engineering process design course has now been harmonized with the NCEES FE/PE exam specification.

First Semester Chemical Process Design I—Theory Meets Practice and Gets Along Well

Part one of the course focuses on an overall approach to design problem solving, supporting engineering and economics calculations as well as Environment, Safety and Occupational Health (ESOH) considerations. Students learn how to analyze externally posed problems (design basis development), create diagrams, i.e., to represent their ideas in flow sheets and to perform basic material and energy balances necessary to size and estimate the cost of requisite process equipment. A chemical engineering accounting system based on the fundamental scientific principles of the conservation of matter and energy is coupled with design computational methodologies to size, select and cost individual and grouped process equipment. Heat transfer, mass transfer, chemical reactions, pumping, turbines and compressors are examples of the technical areas covered along with engineering economics, safety and environmental considerations. NCEES FE and PE specification topics serve as a useful, if not obvious guide to the subject matter as the two together essentially comprise the practice of chemical engineering.

Emphasis in chemical process design is placed on assuring that all raw materials entering a chemical process are accounted for either as finished products or waste by-products, with the difference being permitted environmental emissions. Similarly, energy must be accounted for and used maximally in order to assure project success and to minimize environmental footprint. Methods to maximize raw material conversion to products, while minimizing energy use and waste generation are brought to focus for our students here for the first time as juxtaposed against capital cost considerations e.g., investment, operation in relation to the time value of money.

Real world examples are used whenever possible, drawn from experience over the past 25 years consulting and design engineering. In addition to a textbook, reference material from existing chemical engineering literature is provided to the students as the need arises, as supplemental "toolbox" learning for life material. The aim is to present a systematic approach to tackling engineering design problems along with the development of tools the students can take with them for use beyond the academic portion of their careers. This draws heavily upon the university-wide prerequisite course work exposure with an eye towards translating theory into practice and vice versa. Emphasis is placed on hand (calculator based) or spreadsheet enhanced calculation procedures in preparation for second semester incorporation of computer aided process design. This approach provides students with a method of validating as well as improving model results instead of relying solely upon them.

Second Semester Process Design II—Senior Design Projects with Industry Advisors

Only in existence for a little over a hundred years, the original intent of the new discipline of Chemical Engineering was to systematically and economically turn raw materials into finished products on a large scale. As the chemical process industry matured, the need to protect human health and our environment, conserve diminishing natural resources and in short to promote sustainable engineering design was added.

Accordingly, the first 2 years of second semester design projects focused on biofuels production and chemical manufacturing from renewable resources, respectively. This year's project focus employed a novel synergistic approach namely chemical manufacturing from gasification based raw materials. Gasification can take a wide variety of organic material as feedstock; virtually anything from municipal solid waste (garbage) to otherwise useless cellulosic material or even hazardous waste and convert it into CO and H_2. These two gases (Syngas) can, in turn, be made into a wide variety of finished chemicals that would otherwise require petroleum-based feedstocks. An additional learning component is gained by pairing these facilities together as the overall energy efficiency can go from 60–70% to in excess of 90%—that is if the students find out how to do this—the fun part of the course. Four team pairs were established, with the students having a hand in picking the type of gasifier to design as well as the chemical to be manufactured from the Syngas feedstock.

Each of the eight teams was assigned a practicing industry client/mentor to assist in an advisory capacity and to play the role of commissioning client for whom the process design is being performed. In total this past semester, six working chemical design engineers served as client/mentors, five of whom are licensed professional engineers. The student teams were formed by the course coordinator to assure a balance of strong first semester Design I performers on each Design II team. Student leaders, chosen by the team members, watch over team progress and deliverables as well as individual participation levels.

The Design II focus is on project development of conceptual process design to bring decision points to go/no go based on real-world project engineering considerations that review technical feasibility, address ESOH considerations, and evaluate all economic aspects. A layered approach is taken, namely the entire project report is laid out in advance and is worked on at all times. All eight groups meet every 3 weeks for 5 h to present their team progress, each having its own "wiki type website" where all can observe progress. A course coordinator generated timeline is used to set project deliverables for the 5 triweekly meetings with emphasis on a different area of the design report until eventually all reporting elements are filled in. In addition to the in-class course requirements all eight teams participate in two public presentations, namely (1) The Annual UIC College of Engineering Design Expo where all engineering disciplines compete against each other and (2) The Annual American Institute of Chemical Engineers (AIChE) Chicago Section

Student Poster Competition including the three chemical engineering degree granting Chicago-land based universities, Northwestern, Illinois Institute of Technology and UIC. To complement the first semester calculations approach, the students use Aspen Plus (Aspen Technologies), a popular industry computer process design/simulator that incorporates process flow sheeting, equipment sizing and performance as well as cost estimating.

2. *Collaboration of faculty, students, and licensed professional engineers*
The course coordinator, (J.P. Perl) is an Adjunct Professor and a PE working full time as a consulting engineer. During the second semester, industrial mentors come together with the course coordinator to review the existing Design II course format and to select the overarching design topic for the student projects. These mentors, five out of six who are PE-licensed helped create the original Design II course structure 3 years ago, based on a challenge put to them at the time, namely, what was desired in new hire educational preparation. These men and women work for UOP-Honeywell, Jacobs Consultancy, Middough Engineers, and Ambitech Engineering, and are all actively engaged in national and international chemical process design projects.

While both semesters are guided by meeting ABET requirements, the NCEES FE/PE exam specification serves somewhat as a teaching template that covers the basic practice of chemical engineering. While it isn't possible to cover all of ChE in the time allotted, the basic approach to engineering and design problem solving can be laid out. The American Institute of Chemical Engineers (AIChE) strongly supports licensure and populates the 30-person volunteer Chemical Exam Committee working at least twice yearly with NCEES to oversee the creation and maintenance of the ChE discipline exam, and from time to time to reevaluate the body of knowledge. Exam committee members are drawn from industry, academia, and government, representing the gamut of chemical professional engineering practice. The course coordinator is a member of this committee and this interaction has also helped shaped the development of this course.

The eight student design teams are arranged in four gasifier/chemical producer team pairs. The chemical production team sets the specification of Syngas that the gasifier team must meet. This involves a degree of interaction and economic balancing that simulates real-world considerations, e.g., facility siting, raw material supply and transportation amongst others. Every 3 weeks the teams each make 30-min project progress presentations, 20-min talk, 10-min Q&A, to the entire class and mentors. During this 5-h joint session, they receive feedback from all the mentors as well as their classmates. In this manner, they gain valuable experience on their feet defending and correcting their ideas—a time tested system common to the ChE design community that serves to optimize design as well as maximize safety. It is enjoyable to see how they progress from largely theoretical approach to learning, to one that attempts to demonstrate how to temper theory with practice. The 5 tri-weekly progress meetings are graded, feedback is provided to students and each group moves on to the next phase of design and must address earlier deficiencies identified during the preceding 5-h session. Quizzes, midterm, and final exams are used to gauge individual progress and comprehension of lecture material and count for 25% of the class.

In between the triweekly progress meetings, a lecture series has been developed to support the general design approach methodology. The teams also have weekly meetings with their mentors. The mentor dedication is incredible and key to the success of Design II. They each come for 6 h every 3 weeks on top of the other meetings and this is the real secret to success in my mind. And so during the semester, they donate an additional 60 contact hours with the students, 30 of which require they be away from work (all the companies support this!) as well as an additional 40 h of planning.

Examples of supplemental mentor lectures to the entire class include "How to Make Presentations," "What Should Be in The Report," "Fundamental Engineering Economics Including Spreadsheet Template for Project Review," "Use of Aspen Technologies Aspen Plus Design Program for Economic Evaluation," "The OSHA Process Safety Management Regulation," "The Environment, Safety and Occupational Health Regulatory System of the United States," "How to Conduct Engineering Technical Literature Research," "Plant Layout," "Plant Utilities and Offsites," "Creation of Piping and Instrumentation Drawings (P&ID)," and "Design/Build Project Execution."

3. *Benefit to public health, safety, and welfare*

Only in existence for a little over a hundred years, the original intent of the new discipline of Chemical Engineering was to systematically and economically turn raw materials into finished products on a large scale. As the chemical process industry matured, the need to protect human health and our environment, conserve diminishing natural resources, and in short to promote sustainable engineering was added.

Once built, chemical manufacturing plants of all kinds are difficult and expensive to change. Now more than ever greater emphasis must be placed on doing it correctly the first time. This two-semester course provides a sharp focus on minimizing raw material and energy usage. Incorporating industry and AIChE developed safety training to protect human operators and surrounding communities from accidental and/or unnecessary environmental release and exposure.

As we are running out of natural resources in the face of ever expanding populations, sustainable engineering has been the catch phrase of the twenty-first century for chemical engineers. This year's project focus employed a novel synergistic approach namely chemical manufacturing from gasification based raw materials, a process that can take garbage and turn it into useful products as well as surplus energy. These projects all represent a significant carbon footprint reduction through increased energy efficiency.

4. Multidiscipline and/or allied profession participation

Chemical process design engineering involves interaction with all major engineering disciplines and the trades. The course revolves around the engineering and construction industry (E&C) as well as plant operations experience of the providers. The chemical process design engineer must understand a variety of ancillary topics and so the course also emphasizes: mechanical equipment, e.g., compressors, boilers, understanding ASME pressure stamps, and NEC and NFPA electrical and fire and explosion proof requirements as well as other industry standards. Transportation requirements as well as plant layout and siting, areal civil/structural engineering, constructability and maintenance access considerations for design/build are emphasized. Interaction with equipment vendors as well as facilities already in operation or in demonstration plant design phase is also encouraged especially as pertains to sizing, costing and project validation. The outside professional community has been most helpful in this arena.

Senior design students are encouraged to attend local AIChE dinner meetings ($5 versus $35) where they are exposed to interdisciplinary presentations such as power generation, safety, USEPA and OSHA regulations and other topics of broad interest to local professional members. A yearly poster session is arranged by AIChE where they can present their projects, compete for prizes, and receive professional feedback.

5. Knowledge or skills gained

Upon completion of this two-semester course, our students have developed initial computational skills, learned how to pose problems (design basis) and how to draw out, size, and cost typical projects. Because of the course format employed, each design team gains close familiarity with the seven other ongoing projects. Working knowledge of Aspen Plus, an industry standard design and economic estimating program, is also gained. All the students gain experience and confidence through presenting their projects five times in a Design class/mentor settings and two times in an external peer review setting. The team approach assures that the basic elements of a unique chemical process design is created in an industry recognizable format that all students can learn from and take with them as a template for professional engineering life upon graduation. A complete syllabus follows showing detailed exposure of students to process engineering design.

6. Detailed Course Structure

The following syllabus is provided to the students and is used as a metric and grading scheme throughout the course. This year's gasification/chemical production topic is used:

The course is laid out as follows:

1. Five Specific Milestones spaced every 3 weeks with group presentations.
2. Lectures interspersed to support Design Project Objectives.
3. Quizzes, Midterm, and Final Based on Lectures and general design principles.
4. Five Group meetings every third Tuesday starting third Tuesday of semester.
5. Lectures by Course Coordinator and Mentors and Others.
6. AIChE University Night Dinner Meeting—Student Poster Session Competition.
7. UIC College of Engineering Design Expo 2011—participation required for graduation.
8. FE Exam April 2011—Optional in lieu of final exam.

Additional guidance to students for this semester:

The course will build on the fall semester design principles and focus on Gasifier-based chemical production. Feedstocks may include coal, biomass, or municipal solid waste (MSW). Products include: ammonia, chemical waxes, formic acid, dimethyl carbamate, ethylene glycol, acetic anhydride, vinyl acetate, olefins/aromatics, aldehydes/alcohols, acetaldehyde/ethanol, and others. Processes can be selected from DOE or EPA websites, other societies, existing and startup company websites. Thermal, Bio, or catalytic processes or anything that does not violate laws of physics are allowable. See also book "Gasification" by Higman and van der Brodt. Processes must be feasible or demonstrated, i.e., chemistry exists.

Course Grading

- Projects will count as 75% of the grade.
- Midterm—10%, Final—10%, Quizzes—5%, and Attendance—Swing Vote.
- Teams will select a leader who maintains project timeline and individual participation %.
- Course Coordinator to Grade Exams.
- Coordinator and Mentors Grade Projects.
  - Graded after each group presentations.
  - Advisors grade all projects.

General Grading Categories—Reviewed after each of 5 triweekly presentations and at course end. Specific grade categories are for each of the 5 triweekly presentations 1–4, with 4 excellent and 1—unacceptable.

The presentations are scored as a group in the following general categories:

1. Participation.
2. Report—Written.
3. Report—Presentation.
4. Level of Detail.
5. Satisfied Overall Objectives.
6. Progress Adequacy.

In addition, as the project advances every 3 weeks, new specific technical report elements are graded as well, in order to keep things on track and to deficiencies to improve for next meeting.

The Course Coordinator converts these scores into letter grades at a semester's end that also includes written exams and attendance.

Design Project Deliverables

The overall Design II course objective is to produce a written and oral project report: ± 30–50% process design cost estimate (Conceptual Design—Stage Gate 1 Level)

Final Report to include:

Executive Summary
Discussion
Recommendations

Appendices

1. Design Basis
2. Block Flow Diagram
3. Process Flow Diagram showing major equipment
4. Material and Energy Balance including Integration Review
5. Calculations
6. Annotated Equipment List (Data Sheet)
7. Economic Evaluation factored from Equipment Costs
8. Utilities
9. Conceptual Control Scheme
10. General Arrangement—Major Equipment Layout
11. Distribution and End-use Issues review
12. Constraints Review
13. Applicable Standards
14. Project Communications File
15. Information Sources and References

Timeline of Course—Five Group meetings spaced 3 weeks apart are used to pace project development and acclimate students to meeting project deadlines and incorporating midcourse correction and refinement developed through open critique and peer review. Quizzes and exams are interspersed during and at the end of the course.

Meeting 1—Pose and Set Team Problems

- Design Basis
- Competing Processes
- Conceptual Proc Block Flow
- Report Outlines Established

Meeting 2—Preliminary Design Presentations

- Flow Sheets
- Material and Energy Balances (Identify users and sinks)
- Data (Design basis, sources for data and, operating conditions and assumptions)
- Hand Calculations (Mathcad, Solver, etc., excel, but not Aspen yet) at least one hand calculation
- Rough Economics (cost, incl. Labor/equipment, sales, energy, first pass to allow changes)

Meeting 3—Presentations in Aspen Plus computer-based design and costing software

- Flow Sheeting
- Energy Sinks/Loads and Streams Identified Stream numbers and components etc. P&T
- Equipment Sizing
- Costing using the ASPEN Process Estimator Software

Meeting 4—Final Design and Engineering Review

- Process Flow Diagram Complete
- Initial Control Scheme
- Plant Layout (General Arrangements)
- Calculations
- Refined Economics
- Report

AIChE Annual Student Poster Competition

- Northwestern, Illinois Institute of Technology and UIC
  UIC College of Engineering Design Expo

- All Departments Compete
- Awards Based on Judging
- Participation Required for All Disciplines to Graduate

Meeting 5—Final Report Presentation and Submittal

- Written Report for Grading
- Invited Guests Review

Final grading of the projects is done in group session with the Course Coordinator and Mentors. The course coordinator then combines this with exam scores to translate to student letter grades.

Final Thoughts on the Design Course

During the next 3 years, the course built on the combination of integrated power and chemical production from gasification. Ripped from the headlines, this was inspired by the surge in fracking with resultant increase in natural gas (NG) production along with concurrent NG price reduction. By locating our student projects in Williston ND, an additional element of realty was introduced that of the need to locate near resources. It should also be noted that the literature from other engineering firms has also started to reflect this integrated approach.

Appendix C
Teaching Senior Design Examples

C1.1 Teaching Designing a Combined Power and Chemical Production Facility—Sample Outline

Note—The following outline is presented from the Engineering Process Design course I taught for 6 years at the University of Illinois at Chicago (UIC) [1]. So, why present this here?

1. It is difficult to get proprietary process detail like this from industry.
2. I was assisted by 6–10 Chicago area design engineer colleagues in devising a complex, integrated power and chemical production facility.
3. As will be seen later, practice-based design teaching is of paramount importance if we are to inculcate our young men and women engineers of the future with SE design principles, a goal which my fellow industrial engineers heartily agree with.
4. My academic colleagues may find this useful in constructing their own senior capstone design course.
5. My junior engineering colleagues may find this a useful guide.
6. My senior engineering design colleagues will fill my mailbox with numerous suggestions for improvement!
7. Managers can use this to "Get Smart" on Sustainability Engineering.

I have left this in syllabus outline form as it was developed, as much as a design checklist as anything else. The two block diagrams are the final result of interconnecting ten different chemical processing blocks along with power generation to create multiple merchant products. The corn to chemicals example comes closest to sustainable as everything is made from corn, nature's own renewable resource. Later we will look at an example of Frack Gas to Chemical Products which of course is not as sustainable in the pure sense of the word as natural gas is a finite

J. P. Perl, *Sustainability Engineering for Enhanced Process Design and Manufacturing Profitability*, https://doi.org/10.1007/978-3-031-52363-2

resource. Still it too represents the manner in which disparate production elements can be combined with power generation that was needed anyway, in an extremely efficient manner.

✈- An unexpected benefit of SE design is localized power generation. This comes at a most opportune time for the power industry that is near capacity and is already using consumer efficiency (Energy Star–based) improvements as a cost avoidance tool, thereby removing the necessity of installing new capacity. So, SE could not have come at a better time! ✈-

C1.2 A Practice-Based Approach to Teaching Chemical Process Design: Sustainability Engineering

I taught the two-semester senior design course at UIC for 6 years. While the basic principles of ChE design are well documented elsewhere, how to think creatively to enhance the design activity is the approach taken with my young charge-lings. During the first semester, I introduce basic principles of process design much like a toolbox as outlined in the preceding table. I present design tools, tips, and techniques with emphasis on incomplete problem statements to encourage early identification of known unknowns and unknown unknowns (Rumsfeld's Rules). Most of the first-semester examples came direct from my practice, often in real time, albeit sanitized to omit confidential information. During the second semester, I was assisted by as many as ten of my fellow Chicago area process design engineering practitioners, each with an average of 35 years of design experience to mentor teams of four to five students (Class size of 30–40 per semester). The mentors and I created the second semester problem with the objective of teaching the integrated process and power methodology described herein coupled with SE principles, one that optimizes material and energy use by coupling related as well as unrelated processes and power generation for internal and external use. Emphasis is placed on sustainable approaches. In one project, one student team designed an algae-to-oil to chemicals project and were quite surprised at the vast amount of land required for this solar-based technology. They soon learned this unsustainable approach would need a very good chemical reactor design to reduce the footprint by 2–3 orders of magnitude in order to approach sustainability. (This is actually being worked on professionally as I write this). This is a big challenge, but planet earth has done quite well with photosynthesis thus far. (Note—One of the teams designed an Algae to Oil process = CO_2 sequestration done properly through consumption and not burial!)

During the second semester, students are exposed to a real-world, practice-based design problem that challenges them to approach an integrated, sustainable design to minimize raw material and energy utilization while maximizing safety and profitability ... all in a sustainable manner. Many of the problems were borrowed from the literature or created out of whole cloth. At this nuts and bolts stage-gate decision-making level, it was doubtful that the young design engineers would be able to have too much to say about anything SE at first, but they often surprised us by identifying and developing options on their own. They certainly will, however, as they progress through their careers and this classroom exposure is meant to start the thought process going of always questioning why they are doing something and the effects of those decisions beyond their immediate event horizon.

Class progress is measured by five triweekly team presentations taking from 4 to 6 h where all students and mentors alike provided continuous improvement style input—just as it is done in industrial settings.

C1.3 Design Educational Standards

The National Council of Examiners for Engineering and Surveying (NCEES) (Appendix B) is a national nonprofit organization dedicated to advancing professional licensure for engineers and surveyors. It develops, administers, and scores the examinations used for engineering and surveying licensure in the United States. Based on my participation on the NCEES exam development committee starting in 2007, I believe the ChE PE exam specification itself outlines and so describes the *practice* of chemical engineering. As such I used it to inform my selection and incorporation of the academic course content. Additional details on these standards are found in Appendix B.

A sample syllabus follows that I hope will be useful to both academicians charged with teaching design as well as the numerous practitioners responsible for training new engineers as well. Hopefully I will get feedback on this as well from seasoned practitioners.

So, why look at engineering education in an SE design book? My experience consulting and teaching the senior design engineering course for 6 years as well as serving on the NCEES ChE PE exam committee was quite enlightening. The present ChE baccalaureate academic degree is taught from an ever increasingly research-oriented perspective. I had actually started to think about this after leaving work to return to school for an MS and PhD in ChE. ABET itself recognized the need for practice-based design instruction.

So, when I joined UIC as adjunct faculty, I created an engineering design course to bridge those two worlds by providing a link between theory and practice. I believe strongly in the role of education as opposed to rote training and so emphasis in this 1 year senior capstone course was placed first on basic engineering design calculations and methodology and then on a practice-based real-world overall design problem for the entire class, purposely mentored by engineering practitioners. Learning how to think, particularly in the face of open-ended problems, is emphasized throughout and must always be grounded in sound theoretical basis, informed of course, by practice. The success of practical chemical industrial engineering design whether classic or SE relies on the preparation of our undergraduate engineers, 85–90% of who will rely on this educational foundation when they leave academia and start their industrial careers.

C1.4 Process Versus Product Design

Recently, some schools have added product design to process design (Chap. 1, Ref. [13] and Ref. [18]). Product design is a perfect follow-on to the present undergraduate academic approach based on *Transport Phenomena Textbook* by Byrd, Stewart, and Lightfoot (1960) (Ch2:5). As all new chemical products must go through an EPA review during which time SE design principles will assure positive long-term safety and economics, so of course, SE is applicable to both process and product design. The book by Seider et al. is a thorough academic presentation of this design topic and is in keeping with good SE design principles.

An entire industrial support industry has grown up to serve the CPI. Many new developments from academia as well as private funded R&D enter into the application stream in this manner. These become commercial off-the-shelf technology (COTS) that practitioners can apply. New heat exchangers, bioreactors, genetic methods, etc. assure continuous technology improvement and development. Patents are important to a corporations' financial viability, so the industry–academic R&D interface needs special management, and it is here that academic product design finds a sound home as a natural concomitant to process design.

Reference

1. Perl JP. A two-semester senior chemical engineering capstone design course, University of Illinois—Chicago 2008–2014 inclusive

C1.5 First Semester Fall Syllabus

Aug 26, 2014
Chemical Engineering Design I
University of Illinois—Chicago
ChE 396 Fall 2014
Copyright 2014
By
Professor Jeffery P. Perl, PhD, PE, CHMM
Fall Course Syllabus
Concepts of Chemical Process Design

The adopted course textbook is *Chemical Engineering Design*, by Towler and Sinnott, 2nd edition, 2012. Additional course material that I have found useful in practice will be provided as the subject areas arise. The textbook provides good reference material useful to the present as well as the spring (ChE397) semester, and well into your practice as chemical engineers, whether in industry, academia, or government service. For some reason, the chapters on energy and material balances were eliminated in the new edition, so I have provided them online for your reference.

Prerequisites for fall ChE 396 include: CHE 101, 201, 210, 311, 312 and CHE 313 and CHE 321. During this, your senior year at UIC, you will also take additional ChE courses and all of these are applicable to design. Feel free to examine how these other senior courses apply to design and always ask questions as they arise in your other classes! An engineer should always ask: "How can I use this" when coming across new material. The answers may surprise you and also keeps the senior year interesting!

In general, the ChE396 course content has been chosen to help provide a bridge between theory and practice by taking into account the sum and substance of your first 3 years of study of math, science, general and chemical engineering. Among other things, we will examine the material and energy balance–based application of heat transfer, mass transfer, basic reactions, process thermodynamics, conceptual process control, fluid flow, energy and power use, source and conversion, engineering economics, environmental, health and safety, and sustainability concepts as applied to process design engineering. Emphasis is placed on acquiring professional engineering practice skills including technical, health, safety, and environmental project concerns, as well as a *sustainable* and ethical approach to all that we do as engineers. All students are encouraged to prepare and sit for the Fundamentals of Engineering Exam (FE) offered in October and April each year, and the course curriculum draws heavily from the FE and PE exam specification found at www.ncees.org. Although not a requirement, many of our seniors take this exam. (In lieu of the spring final).

An additional aim of the Fall ChE396 course is to prepare you all for the spring second semester, Design II ChE397, course during which time an industrial practice-based design problem will be attacked by the entire class in team mode. The aim of that problem will be to learn how to prepare a preliminary engineering design—technical and economic evaluation package—with a methodology used by modern engineering design firms. This will be aided, as it has for the past 6 years, by practicing chemical engineering design mentors from the Chicago area, including Ambitech Engineers, Jacobs Consultancy, Middough Engineers, and UOP-Honeywell. An additional useful and exciting close to your UIC experience the College of Engineering's annual spring "Engineering Design Expo" as well as the annual spring AIChE juried poster competition that rotates between UIC, IIT, and NWU.

C1.5.1 Some Basic Rules of the Road

There will be a midterm (40%), final (50%) as well as quizzes or homework (10%). The final exam date/time is set and published by the University. As soon as practicable a midterm date will be selected. Quizzes will be typically announced one session ahead. Classroom attendance is mandatory and records are used to adjudicate borderline performance. All UIC-approved absences, e.g., holidays, religious requirements, and medical, are of course acceptable—please notify Perl. Exam make-ups will be at the discretion of Perl. The spring senior Design Expo and attendance at National AIChE conferences are examples of excused absences. Others are made on a case-by-case basis. Reading assignments aligned with course lectures shall be made from time to time, but students are also expected to use the text as a reference as they would a dictionary or encyclopedia or the library itself. Illustrative design principle problems will be presented and solved during class and a permanent homework assignment is to understand this material as well as quiz and exam solutions. My office is 5–10 min by car from UIC CEB, and I will continue, as I always have, to make myself available by appointment at almost any time (that I am in town). Please call prior to coming to confirm (last minute visits are always acceptable so long as I am in!). Students are encouraged to work together in teams as you assimilate the material. This is not only very useful, but realistic and can be enjoyable as well. Computers and textbooks are allowed on quizzes and exams, smartphones may be used in place of computers, but no communication, verbal, electronic, visual or otherwise is ever allowed at those times. Class is held on Tuesday and Thursday either from 10:00–11:50 am or 1:00–2:50 pm, in room 230 CEB. This often presents a student "refueling challenge" so food/beverages are allowed during class (please dispose of garbage). Develop the habit of coming prepared for class—textbook, notes, calculator, paper, etc.—the basic tools of all self-respecting engineers!

C1.5.2 Nuts and Bolts Design I Course Objectives and Topics

1. To turn raw materials into finished products, usually on a large scale, safely, economically, and sustainably.
2. To troubleshoot chemical production problems, make improvements, and advance our profession.
3. The engineer as "Problem Solver."

Engineering design problems are best addressed starting from the simple to the complex. A clear problem, or, design basis must be verified against an accurate block/process flow diagram. Material and energy balances must all close or be considered lost to the process or worse yet, released to the environment. All major process equipment is listed and sized to assure capacity and proper material of construction. Economics forms the basis of nearly all engineering decisions so your design must be tested against that fundamental measure of success. Safety, health, and environmental considerations must all be engineered into your designs as well as assuring a sustainable supply of raw materials. When complete, your design will be reduced to a set of Piping and Instrumentation Drawings (P&ID). The chemical process design engineer then works with all the other disciplines to accomplish detailed design engineering. Multidimensional layouts are created to establish equipment footprint and spacing to assure safety compliance, operability and maintainability. The final set of IFC drawings will be complete enough for an Engineering & Construction firm to build it and prepare it for startup and operation by the manufacturing client.

In the spring, we will perform a preliminary stage gate 1 design on the selected class problem, or about ±30–50% estimate. This is commonly done to determine the likelihood of project financial success if allowed to go to a more complete level of design ±10%, a sort of go–no-go decision point that requires much attention to detail.

This semester, we will cover the basic topics described below. As they arise in lecture, refer to handouts as well as the assigned text. Specific standing homework assignment is to assimilate the lecture material and past quizzes and exams.

Chemical Engineering Design

- Approach
- Purpose
- Problem scoping.

    Flow sheet and drawings
    Material and energy balances
    Equipment calculations

- Unit Operations
- Reaction Engineering
- Hand Calculations: Heuristics—"Rule Of Thumb"
- Computational Design Methods

    Engineering economics

- Net Present Value, Internal Rate of Return, Depreciation
- Equipment sizing, 6/10 rules
- Total Installed Costs, factored estimates, Simple Breakeven, Stage Gates

    Utilities
    ESOH

- Environment, Safety, and Occupational Health

    Instrumentation, Controls, and Process Safety
    Pilot Studies

- Scale-Up
- Identify Problems

    Theory Versus Practice

- Efficiency
- Fouling
- Scaling
- Plugging
- Mixing
- Mass transfer

Sustainable Engineering

- Design for Environment
- Energy Optimization
- Raw Material Optimization

All-Discipline Approach to Detail Design Engineering
Final Design Engineering and Drawings
Construction, Start-Up, and Operations
Ethics at All Times

C1.5.3 The Bottom Line

When you complete ChE 396, you should be able to state the problem at hand (design basis), draw a process flow sheet (PFD), perform material and energy balances, make equipment design calculations, identify conceptual process control schemes, conduct an economic, environmental, safety, health, and sustainability process review always in an ethical manner, and determine whether your preliminary conceptual process design should go to the next level of detailed design and cost estimate.

C1.6 Second Semester Spring Syllabus

Chemical Engineering Design II (ChE397) University of Illinois at Chicago (UIC)

Spring 2014
Professor Jeffery P. Perl, PhD, PE, CHMM
14 January 2014

C1.7 Corn to Polyols Industrial Complex Example 1

C1.7.1 "Integrated Block Diagram of Sorbitol to PG, EG and Glycerol Industrial Complex"

Syllabus and Course Outline

JPP Note—This year's spring senior design project could not have been possible without the contribution of quite a few Chicago area design professionals, several of whom I have known and worked with over the past 30 years (Corn to Polyols—Here my colleagues Jerry Palmer sketched out wet corn milling process, Dr. Jim Foster of ADM suggested the sorbitol reactor scheme, while Dan Rusinak worked the process integration. More engineering design professional participants can be found in Appendix C). In addition, they were joined by Archer Daniels Midland Decatur, IL HQ Team members who inspired the Corn to Chemicals project. In addition, several team members helped serve as subject matter expert lecturers. They include:

Jacobs Consultancy

1. Dennis O'Brien, PE
2. Bill Keesom
3. Dave Bahr, PE

Middough Engineers

4. Dan Rusinak, PE—Chief Imagination Wizard
5. Patrick Shannon, PE
6. John Micheli, PE

Ambitech Engineers

7. Jerry Palmer—Wet Corn Milling Lead
8. Shannon Brown, PE
9. Tom Stephan

Technical Support

ADM

- Jim Foster—Corn to Polyols
- Chris Hoffman—Tech Advisor

UOP

- Adam Kanyuh—Lecturer

JACOBS

- Laura Weaver—Aspen
- Dan Tong—Aspen

CBI

- Jay Gulotta—Aspen

Project Description

Corn Wet Milling allows for the separation of corn into its component parts, starch, oil, gluten (protein), and fiber. The starch can be dried or can be converted to glucose (also called dextrose) by treatment with the enzymes α-amylase and glucoamylase. The glucose can be used for a number of applications such as fermentation to alcohol, lysine, threonine, etc. Glucose can also be converted to sorbitol by hydrogenation.

Sorbitol can be used "as is" for a number of applications such as toothpaste, sweetener, etc. It can also be used as a feedstock for further conversion to glycerol (also called glycerin), propylene glycol, and ethylene glycol. The reaction between sorbitol and hydrogen takes place at high temperature and pressure over a unique catalyst that produces different mix of products dependent on reaction conditions.

The goal is to determine which mix of products will produce the highest profit stream based on the cost of production and raw material costs. The plant is to process 100,000 metric tons of sorbitol per year.

This semester, you will design an entire plant by breaking it into ten critical pieces. We will have seven four-person teams and three five-person teams to accomplish this otherwise impossible task. In addition, each team will be assigned to a client/mentor from industry with whom to interact (mentor) and please (client).

The financial success of such plants requires that all functional units be operating in linked fashion to provide a continuous, safe, economical production facility. Energy product integration is the key to success, so this is an overall class focus. Chemical products and energy are reused and recycled throughout to make for a sustainable operation. A block flow diagram (BFD) is presented on page 250 followed by tables of initial project data (Page 251). The individual referenced BFD block numbers are shown as (1) thru (9) on the diagram. Safety, Ethics, and Sustainability guides all. The following outlines the course approach.

This year's project was suggested by Dr. Jim Foster, Director of Research and Chris Hofmann, Both of Archer Daniels Midland (ADM), Decatur, Illinois. They were principally responsible also for developing the initial Block Flow Diagram. We are deeply indebted to ADM who has also agreed to serve as technical advisors to our students for the duration. Jerry Palmer of Ambitech Engineers provided detailed BFD's of wet corn milling shown in Blocks 1–5.

(Note: No proprietary information was presented herein. An expired ADM Patent for Sorbitol was used for practice purposes—JPP)

1. Primary Theme—Renewable Resources to Chemicals" Each team shall work on one of the ten operations (Refer to BFD), but the overall objective will be to assure minimum cost, sustainable design in compliance with all Environment, Safety and Occupational Health regulations. All ten blocks *must cooperate* to accomplish the best outcome. The Complex shall be located in Decatur Illinois.
2. Feedstock:
   (a) Corn ← focus for this semester.
3. Data and Resources—nonproprietary information only. (JPP):
   (a) Taken from US Patent, ADM, DOE, and Other sources.
4. Overall project/process to be broken down into pieces:
   (a) Block material balance (MatBal) and energy linked between the units.
   (b) Initial MatBal design basis will be provided.
   (c) Team Leaders meet weekly with MatBal coordinator and Perl to assure convergence.
   (d) Overall process will be "Frozen" within 3–4 weeks (no major changes).

5. Each team will do a complete design around one unit in your project as chosen by your mentors and Perl. Leaders from each group to coordinate stitching together overall MatBal. A MatBal Leader/Perl shall assure convergence. Each team will be responsible for their own feed and production requirements (Design Basis) which the other groups will have to meet. All units are to be fully coordinated regarding Material and Energy. No waste energy or materials is the goal. Weekly team leader meetings with Perl to review issues and balances.
6. Perl will select final team names, members, and client mentors. Input will be taken from first semester ChE396 as well as student requests. You may propose your own teams/leaders subject to approval. An energy, materials and economic czar may be appointed as well.
7. The initial class (Tuesday, 14 January) will set the stage for the course and serve as project kick-off. Any additional updates from your mentors and I will be provided along with a Q&A session for you all.
8. Starting about 1–2 weeks after the first class and occurring every 3 weeks, the first of 5 triweekly project review presentation/reports will occur. These will run from 1 to 6 pm and there shall be no class on Thursday of those weeks. Questions during presentation raised by Perl/Clients/Class shall be addressed ASAP and in no case later than the next progress report, so each team must assign a scribe to take notes. Your report shall be in a state of constant preparation and updating.
9. A mentor/Perl reviewed/approved Midterm Project Progress report of 5–6 pages will be required to take the midterm exam.
10. In between presentations days, teams will meet with mentors by telecon or in person weekly to describe progress. Team leaders shall meet on Tuesdays from 1 to 2 pm with Perl to review coordination and other issues. Notes shall be part of the project report.
11. Teams shall establish Wiki's or equivalent where up to date progress may be viewed by team members, clients, and Perl. Wiki shall serve as the project file and will be graded.
12. Team evaluation will occur during each formal presentation and shall account for 80% of the overall grade. Pop quizzes may be given throughout as well as a midterm (10%) and final exam (10%) all drawn from lectures. Attendance = swing.
13. Those taking the Fundamentals of Engineering (FE) exam in April shall be excused from the final exam. (All FE examinees are responsible to submit their own applications/fees to the State of Illinois in a timely manner to assure a seat for the exam). Attendance will be verified.
14. On April 9, 2014, the AIChE-Chicago Section "University Night" meeting will be held at IIT this year and seniors will enter their projects in the poster competition between UIC/NWU and IIT. Cash prizes are awarded and serve as practice for Design Expo!

15. Students are responsible for entering ChE397 team projects with COE in a timely manner for competition in Engineering Design Expo. Mind your schedules!
16. Final Course Project Presentations shall be made to Mentors and outside reviewers.

Initial Process Details

The Polyols production process is based on US Patent 6,291,725 B1. The attached drawing shall serve as a point of departure for dividing into ten-student teams. Over the break, more detail will be fleshed out and student teams will be assembled by Perl.

Patent Abstract

"The present invention provides a process for preparation of low molecular weight polyols from high molecular weight Polyols in a hydrogenolysis reaction under elevated temperature and hydrogen pressure. The process comprises providing in a reaction mixture the Polyols, a base, and a metal catalyst prepared by depositing a transition metal salt on an inert support, reducing the metal salt to the metal with hydrogen, and passivating the metal with oxygen, and wherein the catalyst is reduced with hydrogen prior to the reaction. In particular, the process provides for the preparation of glycerol, propylene glycol, and ethylene glycol from sugar alcohols such as sorbitol or xylitol. In a preferred process, the metal catalyst comprises ruthenium which is deposited on an alumina, titania, or carbon support, and the dispersion of the ruthenium on the support increases during the hydrogenolysis reaction."

Simplified Process Description

Study the Block Flow Diagram to find all the products and intermediates that are made. Both energy and materials are used in this complex in a sustainable manner to maximize benefit and minimize environmental and health and safety impact. Approximately 7,000,000 bushels/year of corn feed will be delivered by rail car to the complex. The Wet Milling Process is performed in the first 5 blocks to create the various intermediate products shown in the BFD. The interactive material balance is actually quite complex, requiring close design coordination between the blocks and the rest of the complex as well. As the goal of this Basis patent is to produce Polyols from sorbitol, the starch produced is first converted enzymatically to dextrose and then hydrogenated to sorbitol. The sorbitol is fed at 100,000 metric tons/year, to the polyols reactor. The reactor products must then be separated and concentrated to saleable merchant products EG, PG, and Glycerol. C1–C4 alcohols are separated in the recycle block which also serves as a material balance unit and environmental control system, including sulfur production. A combined heat and power block will manage fuel, steam, electricity, hydrogen and serve as an energy balance unit (also shown as Fig. 8.7).

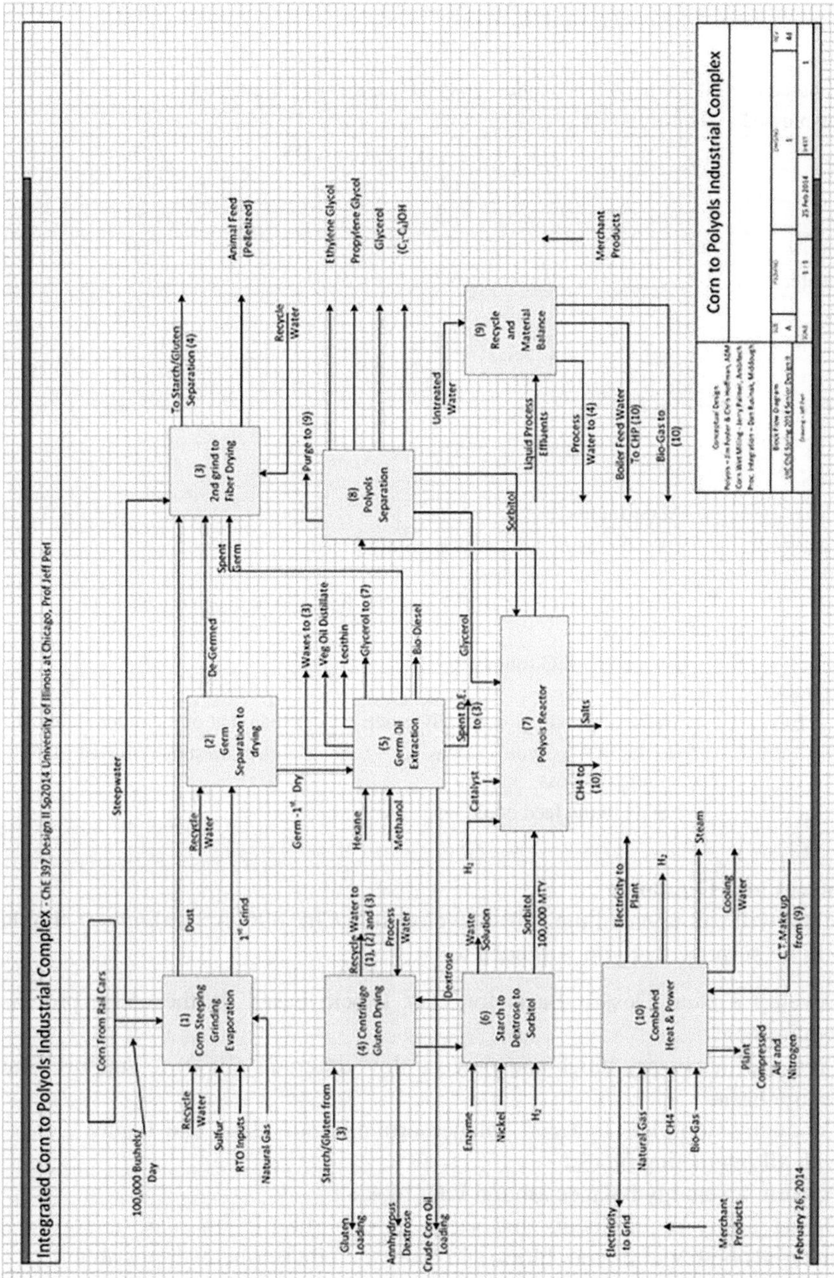
Integrated Corn to Polyols Industrial Complex - ChE 397 Design II Sp2014 University of Illinois at Chicago, Prof Jeff Perl
Corn From Rail Cars
100,000 Bushels/Day
(1) Corn Steeping Grinding Evaporation
Recycle Water
Sulfur
RTO Inputs
Natural Gas
Steepwater
Dust
1st Grind
(2) Germ Separation to drying
De-Germed
(3) 2nd grind to Fiber Drying
To Starch/Gluten Separation (4)
Animal Feed (Pelletized)
Spent Germ
Germ - 1st Dry
(5) Germ Oil Extraction
Hexane
Methanol
Waxes to (3)
Veg Oil Distillate
Lecithin
Glycerol to (7)
Bio-Diesel
Spent O.E. to (3)
Purge to (9)
(8) Polyols Separation
Ethylene Glycol
Propylene Glycol
Glycerol
Starch/Gluten from (3)
(4) Centrifuge Gluten Drying
Recycle Water to (1), (2) and (3)
Process Water
Gluten Loading
Anhydrous Dextrose
Crude Corn Oil Loading
Dextrose
(6) Starch to Dextrose to Sorbitol
Enzyme
Nickel
Waste Solution
Sorbitol 100,000 MTY
Catalyst
Glycerol
(7) Polyols Reactor
CH4 to (10)
Salts
Sorbitol
Untreated Water
(9) Recycle and Material Balance
Liquid Process Effluents
Process Water to (4)
Boiler Feed Water To CHP (10)
Bio-Gas to (10)
Merchant Products
(10) Combined Heat & Power
Electricity to Plant
Steam
Cooling Water
C.T. Make up from (9)
Natural Gas
CH4
Bio-Gas
Plant Compressed Air and Nitrogen
Electricity to Grid
February 26, 2014
Corn to Polyols Industrial Complex

Data and Design Basis Requirements
Some Important Corn Properties

Some Important Process Requirements

Yields based on 100 lbs. of corn dry solids, 56 lbs./bu. corn at 15% moisture		
All # expressed as # of dry solids		
	# per 100# corn DS	#/Bu
Fiber	12.7	5.99
Germ	7.5	3.54
Gluten	5.5	2.59
Steepwater	6.7	3.16
Starch	66.7	31.46
Corn cleanings	1.802	0.85
Total		47.60
Total DS in a bushel of corn		47.60

Other	ADM Inputs
Corn Wet Mill Energy	100,000 metric ton per year sorbitol feed
150 PSI	The combined yield of EG and PG should be at least 60 wt% to be economical
5 PSI for evaporators	Typical PG–EG ratios are ~2:1
Fuel—TBD	Methanol production usually much greater than that of the other alcohols
	Higher MW polyols such as 1,2,5,6-hexanetetrol must be removed in the refining process
	NaOH in the feed could be as much as 1%

Resources and Guidance
This class will call upon all the skills developed in ChE 396. Accordingly it should come as no surprise that you will need to:

1. Develop a clear Design Basis for your Block, based on the entire process requirement
2. Create all necessary block and process flow diagrams (PFD) to support your design basis
3. Conduct a complete material balance around your block and assure closure with all interacting blocks
4. Perform an energy balance around your block
5. Define utility requirements and seek out opportunities to share energy and materials both within your block and cross the others wherever possible
6. Develop a conceptual process control scheme for your block

7. Identify, with your mentor, an area to perform a complete design around:
   (a) Size and select process equipment
   (b) Design pumping system
   (c) Model this in Aspen
      - Equipment
      - Cost
8. Perform a net present value (NPV) engineering economic determination for your block and coordinate this with an overall balance.
9. Overall Material and Energy Balances to be done in Excel with excel solver as opposed to ASPEN.
10. Create Wiki or equivalent and keep up to date project progress.
11. Present a plant layout.
12. Be ready to present progress at 5 triweekly meetings, AIChE Poster Session and UIC COE Engineering Design Expo.

Additional Resources
Lectures during the semester will include economics, report writing, presentation building, P&ID, Utilities and Process Integration, AspenPlus 101 and Economics Package, Plant Layout, Environment, Safety and Occupational Health, Process Safety Management and others.

END Corn to Polyols Student Design Problem

C1.8 Spring 2013 Student Design Problem Example 2

The following is taken from my spring 2013 Chemical Engineering II Design Class. As discussed in the NCEES section, I taught the two-semester design class at the University of Illinois at Chicago for 6 years (2008–2014) as an Adjunct Professor. Just for fun, I brought in six to ten of my Chicago area design engineering professionals to help create the second-semester senior project. This was set up as a sort of project design simulator. The projects were assembled as if they were real client assignments. A combined approach was taken combining disparate processes together in order to minimize energy use and maximize material conversion to merchant products. I have included this as an example as it highlights the approach to sustainability engineering and presents enough detail to illustrate that. I am quite proud we were able to expose our students to real-world engineering design emphasizing team work and environmental responsibility. I have noted pertinent references from the web and elsewhere.

JPP Note—This was culled from bits and pieces of actual as well as proposed processes as referenced at the end of this chapter. No proprietary information was used. This was part of a class syllabus and what remains are essential design elements minus class instructions. A block diagram is included to show both complexity and simplicity (This mythical conceptual design was conjured up by my colleague, Dan Rusinak, as our chief imagination wizard.)

C1.8.1 "Integrated Shale Gas Industrial Complex"

Outline

Shale Gas Production is flourishing in the USA. This has caused a fivefold reduction in the price of natural gas over the past few years leading to a flurry of new US projects to capture, recover, and now use this abundant natural resource for both energy and chemical production. The focus is on sustainable methods to use both the gas and the produced "frac" water.

The entire plant is broken into nine critical pieces with individual design efforts to be coordinated among them. The financial success of such plants requires that all functional units be operating in linked fashion to provide a continuous, safe, economical production facility. Energy product integration is the key to success, so this is an overall class focus. Chemical products and energy are reused and recycled throughout to make for a sustainable operation. A block flow diagram (BFD), spreadsheet schedule of activities as well as the report format as depicted by box 1–7 is attached. Safety, ethics, and sustainability guides all. The following outlines the course approach.

1. Primary Theme—Integrated Shale Gas Industrial Complex. Each team shall work on one of the following (refer to BFD), but the overall objective will be to assure minimum cost, sustainable design in compliance with all Environment, Safety, and Occupational Health regulations:
   (a) Shale Gas Treatment Plant.
   (b) Ammonia (NH_3) Plant.
   (c) Fertilizer (UAN-32) Plant.
   (d) Combined Heat and Power (CHP) Plant (energy balance point).
   (e) Air Separations Unit (ASU) and Syngas Plant.
   (f) Fischer-Tropsch Liquids (FTL) Refining Plant.
   (g) Direct Iron Reduction (DRI) Plant.
   (h) Material Balance and Process Effluent (MBPE) Upgrade Plant (material balance point).
2. Feedstock.
   (a) Natural Shale Gas ← focus for this semester.
3. Data and Resources.
   (a) Taken from USDOE and Other sources.
4. Overall project/process to be broken down into pieces.
   (a) Material balance (MatBal) and Energy linked between the units.
   (b) Initial MatBal design basis will be provided.
   (c) Team Leaders meet weekly with MatBal coordinator and Perl to assure convergence.
   (d) Overall process will be "Frozen" within 3–4 weeks.

Project Conceptual Design Basis

The overall objective of this plant is to integrate the production of various chemicals and iron by use of natural gas from shale oil. Nine separate units have been identified which either produce merchant saleable products, or intermediates necessary for their production. Collocation of these units will allow for high as well as low quality energy recovery, to minimize environmental impact, improve sustainability as well as enhance safety of operation. As usual, make sure you have a proper picture of your process. Learn your chemistries and perform material balances. Then energy balances. Now look to match up your process with the one before and after or elsewhere as needed. Weekly meetings with Perl will keep on target and refine these goals. Process Safety Management (PSM) shall be emphasized throughout, along with general Environment, Safety and Occupational Health (ESOH) compliance and sustainable design engineering.

Shale Gas Treatment Plant—This plant shall receive shale gas at a rate not to exceed 500 MMSCFD. This gas can be used to create various products by simple separation, e.g., LNG, NGL as well as pipeline quality gas for home and industrial end use. This plant shall focus primarily on pipeline gas for homes and industry, ethane, and CH_4 for the production of ammonia at the rate of 1000 TPD NH_3.

NH_3Plant—This unit shall employ a steam–methane reformer to create syngas, which when combined with nitrogen shall create ammonia. The nitrogen shall be obtained at a purity of not less than 90% from the Air Separations unit. 1000 TPD NH_3 shall be made to be used solely in the manufacture of the fertilizer UAN-32.

Urea Ammonium Nitrate Fertilizer Plant—This unit shall produce UAN-32 fertilizer using the StamiCarbon process by creating ammonium nitrate and urea from 1000 TPD $NH_3$and then combining them along with water to create this (32%, wt, Nitrogen) widely used fertilizer.

CHP Plant—All energy will be processed through CHP which will feature a steam turbine for the generation of electricity, boiler(s) for the production of various steam pressure, low, medium and high as required. CHP shall receive recycled frac water from ***block 9** and upgrade to boiler feed water (BFW) purity required for turbine operation. CHP shall provide process air for compressors, pumps and instrumentation systems as well as plant-wide cooling and process water. All plant electrical requirements shall be met as well as 50 MW electric to the grid. Make-up water shall be taken from area wells, demineralized, and filtered as required.

Air Separations and Syngas Plant—The ASU shall provide N_2 and O_2at purity of at least 90% by volume for NH_3 production and O_2 in purity required for other processes. An auto thermal reformer (ATR) unit shall be designed to create syngas from CH_4 to be used in the Fischer Tropsch Liquids Plant as well as the Methane to Gas Plant. This unit will also separate for downstream reuse of excess CO and CO_2.

Fischer Tropsch Liquids (FTL) Plant—The FTL shall be designed to produce liquid products at a rate of approximately 5000 bpd of hydrocarbons. A Medal membrane process shall be employed to adjust the ATR H_2–CO ratio suitable for FTL production. Excess H_2 and product H_2O shall be returned for reuse within the plant.

Direct Fe Reduction (DRI) Hot Briquette Iron (HBI) Plant—This plant shall be designed around the MIDREX process, employing a dry reformer and energy integration to reduce iron ore $Fe_2O_3$to hot iron briquettes for immediate transport into an adjacent mini-mill. Excess CO and CO_2 shall be upgraded as required for use in other plant units.

Syngas to Liquids (SGTL) Plant—Many different products can be made here. This plant shall be designed to accept ATR syngas for the manufacture of propylene for merchant sale.

Material Balance and Process Effluent (MBPE) Upgrade Plant—Process vents and wastewater will be sent here for recovery or reuse based on concentration/type of material. All shale gas produced Frac water shall be processed and sent to CHP. Tank vents and production off-gas shall be processed and recycled for reuse or used as fuel source where applicable. (Actually this block used to be called environmental, but I felt this implied waste, so we turned it into a SE way to get materials and energy back into the process or somewhere we could squeeze more value from it!)

Economic optimization is performed on individual blocks, then on the over-all process. Some will make money, some will lose. Only the overall process viability matters!

JPP Note—Here is the block diagram initially provided to the students. Each of the nine blocks had four to five students and a professional industrial design engineer to serve in the role of client/mentor. A series of lectures were also provided as well as 5 triweekly meetings where all teams presented progress and took criticism to build on for next time. The student teams had were responsible for material and energy balances of their own blocks. They also had to interact between the blocks to optimize energy and mass utilization. I created a material and energy czar to help coordinate this as well as an overall economic czar to combine the economics. Every week I met with the nine team leaders to keep things on track. This 3 semester hour class really was worth 5–6 h, but there was little complaint… unless you consider complaints from the faculty that the seniors were not paying attention in their classes!

The mythical plant was located in Williston ND of course, close to the abundant supply of natural gas. Notice the design produces all the electricity and natural gas needed as well as serving the "Grid" of with excess natural gas as well as electricity. From this BFD, it is apparent where the economy of scale comes into play and very little energy is lost to the environment. Even the iron briquets are fed to an adjacent mini-steel mill (not shown). As the class progressed, a complete material and energy balance along with stage gate 1 economics was created.

This real-world exercise was based on an amalgam of approaches used by modern design firms, with the possible exception that I probably had a bit more control over the class as their grades depended on it! This class was enjoyable for all as the students developed a much better idea of what would be expected from them after leaving theory land. Also developed confidence in that all learned how to present not only their own team's work, but mastered the entire project. Hey that's real life.

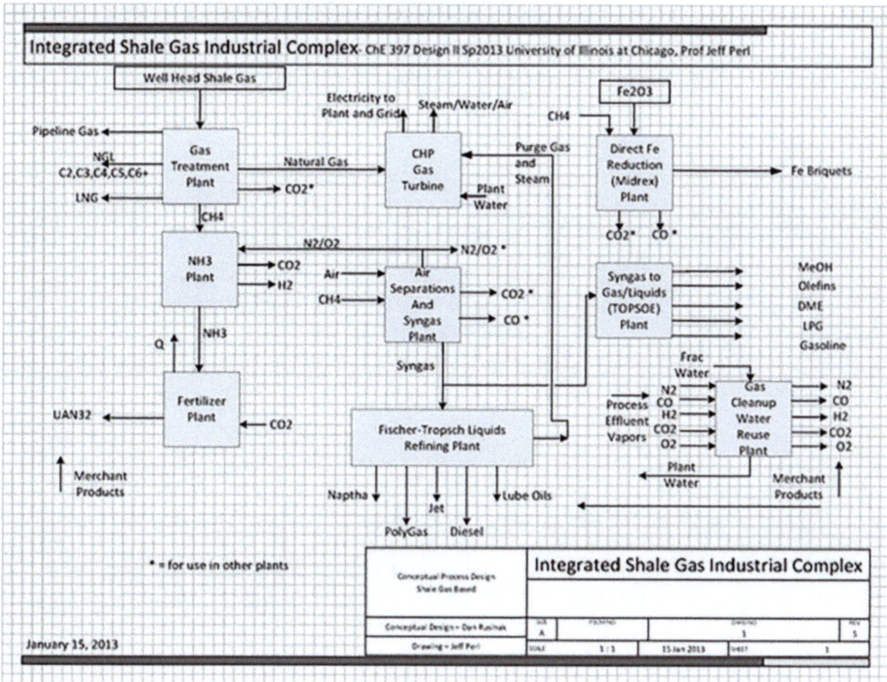
Integrated Shale Gas Industrial Complex- ChE 397 Design II Sp2013 University of Illinois at Chicago, Prof Jeff Perl
Well Head Shale Gas
Pipeline Gas
NGL
C2,C3,C4,C5,C6+
LNG
Gas Treatment Plant
Natural Gas
CO2*
CH4
Electricity to Plant and Grid
Steam/Water/Air
CHP Gas Turbine
Purge Gas and Steam
Plant Water
Fe2O3
CH4
Direct Fe Reduction (Midrex) Plant
Fe Briquets
CO2* CO*
NH3 Plant
N2/O2
N2/O2 *
CO2
H2
Air
CH4
Air Separations And Syngas Plant
CO2 *
CO *
Syngas to Gas/Liquids (TOPSOE) Plant
MeOH
Olefins
DME
LPG
Gasoline
Q
NH3
Syngas
Frac Water
Fertilizer Plant
UAN32
CO2
Fischer-Tropsch Liquids Refining Plant
Process Effluent Vapors
N2 CO H2 CO2 O2
Gas Cleanup Water Reuse Plant
N2 CO H2 CO2 O2
Plant Water
Merchant Products
Merchant Products
Naptha
Jet
Lube Oils
PolyGas
Diesel
* = for use in other plants
Integrated Shale Gas Industrial Complex
Conceptual Process Design
Shale Gas Based
Drawing – Jeff Perl
A
1
5
1 : 1
15 Jan 2013
1
January 15, 2013

Index

J. P. Perl, *Sustainability Engineering for Enhanced Process Design and Manufacturing Profitability*, https://doi.org/10.1007/978-3-031-52363-2

D

MIX
Papier aus verantwortungsvollen Quellen
Paper from responsible sources
FSC® C105338

If you have any concerns about our products,
you can contact us on
ProductSafety@springernature.com

In case Publisher is established outside the EU,
the EU authorized representative is:
Springer Nature Customer Service Center GmbH
Europaplatz 3, 69115 Heidelberg, Germany

Printed by Libri Plureos GmbH
in Hamburg, Germany